BATTLE
TACTICS
of
THE
CIVIL WAR

BATTLE
TACTICS
of
THE
CIVIL WAR

Paddy Griffith

YALE UNIVERSITY PRESS
New Haven and London

First published in the United Kingdom 1987 by The Crowood Press
under the title *Rally Once Again*:
Published in the United States 1989 by Yale University Press.

Printed in Great Britain
by Redwood Burn, Trowbridge

Library of Congress catalog card number: 88-50664
International standard book number: 0-300-04247-7

10 9 8 7 6 5 4 3 2

For Steve Fratt, Joe Park and Gunther Rothenburg,
who started the whole thing

CONTENTS

PREFACE 9

INTRODUCTION
The Alleged Origins of Modern Battle 15

1 The Armies Learn to Fight 29
2 Command and Control 53
3 The Rifle 73
4 Drill 91
5 The Battlefield and Its Fortifications 117
6 The Infantry Firefight 137
7 Artillery 165
8 Cavalry 179

CONCLUSION
The Last Napoleonic War 189

APPENDIX I
The Art of Tactical Snippeting 193

APPENDIX II
The Decisiveness of Civil War Battles 197

NOTES 201

BIBLIOGRAPHY 221

INDEX 231

TABLES

3.1   Union Infantry Armament at Gettysburg

3.2   Armament in the Army of Tennessee, August 1863–
        June 1864

3.3   Small Arms Procurement by Central Governments

3.4   Approximate Costs of Civil War Weapons

3.5   Drill Movements to Load and Fire One Round

4.1   Mobilisation of Regiments in the Civil War

6.1   Ranges of Musketry Fire

AII.1   Military Results of Sample Battles, 1800-1918

PREFACE

This is a book about the way battles were fought in the Civil War – the way soldiers squared up to each other, used their weapons and reacted to the dangers. It is my attempt to correct some of the misconceptions which surround the subject, and to look at familiar explanations with new eyes. By reinterpreting what happened at the lowest levels of tactics, moreover, I hope to find some keys to the way the war as a whole was fought.

The past, of course, is a foreign country, and every historian is to some extent a tourist looking in from the outside upon the people and events he describes. In my own case, I have an added perspective as a British citizen who has literally been a tourist to many of the battlefields on which the Civil War was fought. Since most of the work on this war has been done by Americans, I am hopeful that my view from across the Atlantic will have some new highlights to offer.[1]

Both the tourist and the historian have a duty to be clear about their motives, especially in a case like the Civil War which remains important in modern-day life. We must remember that not all the social and constitutional wrangles of 1861 were laid to rest at Appomattox in 1865. As an outsider to American society I must therefore hasten to plead a strict neutrality in these debates, although for what it is worth I would today classify myself as a supporter of Negro emancipation, states' rights, and almost any lost cause. Against this I would wish to condemn coercion, carpet-bagging and firing the first shot. Sherman's doctrine of warfare against civilians I regard as one of the more vicious military theories of modern times, although it pales almost into benevolence when set beside certain Confederate plans for the massacre of the Apaches[2] or some of their raiders' depredations in the guerrilla war. Yet on the other hand I cannot remain unmoved by the rousing chords of 'Rally Round the Flag' and the rest of the Civil War songs; and as a student of military affairs I stand awestruck before such feats as Lee's manoeuvres at Chancellorsville or the Union breakthrough on Missionary Ridge. Finally, as a loyal Liverpudlian I am mightily tempted to applaud the epic voyage from Merseyside of the *CSS Alabama*, and as a Lancastrian I find I must just as heartily deplore the Lancashire cotton famine which resulted from the Union blockade.

Let the bemused reader attempt to make what he will of my ill-assorted twentieth-century prejudices – they will scarcely help him to understand my motives for writing this book. Much more to the point, perhaps, may be the fact that I am a member of the Vietnam generation, that

9

demographic cohort which attained military age during the period of the Second Indochina War. This experience made the American way of war and its historical background seem suddenly to be a matter of rather more than purely academic interest. One felt that if only one could probe the American military past to find how the patterns were originally formed, and how the various conflicting lessons were digested, one might arrive at a rather better appreciation of the US Army today.[3] The Civil War was clearly a vital turning-point in this evolution, since it was then that American soldiers first shook themselves free from imported military doctrines, and for the first time tackled the problems of warfare on a genuinely large scale. The Civil War also produced almost twice as many American casualties as any other war ever, so the impact which it made on the military mind was presumably proportionately profound.

The experience of attaining military age also seems to have left me with another legacy, for like many another before me I became fascinated to discover just what a battle is, or was, like – preferably without actually venturing into one in person. I suspect that this somewhat unhealthy obsession is really quite common among military historians, and it can even be seen as a precondition of their calling. Each generation has had its own group of military writers who have wished to see the elephant of warfare without getting too close to it, and then relay their findings faithfully and truly to their readers.

It so happens that one of the most successful of all attempts to see the elephant in a war book was made in a novel about the American Civil War – about the battle of Chancellorsville, to be precise. This was of course Stephen Crane's *The Red Badge of Courage*, first published in 1894 when the author was only twenty-three years of age. It would be difficult to overstate the importance of this work to the general development of war writing, since its influence has been enormous and its perceptions remain as fresh and vigorous today as when they were originally penned.[4] The amateur elephant watcher is especially captivated by what Crane had to say, and is led on inevitably into a closer examination of the Civil War battles. There is a sense in which Crane has consecrated these particular combats for every student of battles. I, for one, freely acknowledge my debt to him in much of what follows.

At a more mundane and technical level, I find that another part of my interest in the Civil War stems from my long-standing study of nineteenth-century European combat methods and doctrines, especially those of the French. In the past I sometimes found it difficult to square these with what I knew of the American experience,[5] so I am hoping in the present volume to reconcile the two at last. For example, I have been particularly interested to discover that the French works on the Napoleonic Wars which appeared soon after 1815 were apparently a central influence in the military education of certain Civil War commanders. My enquiry into the Civil War thus arises partly from a feeling that far from being the first modern war (as it is conventionally portrayed), it was in fact the last Napoleonic war. The Civil War can certainly shed a great deal of light upon the earlier struggle, whereas one remains highly sceptical that it can significantly enhance our understanding of the two world wars of the present century.

For the student of Napoleonic tactics, accustomed to a thin and insubstantial diet of secondary sources and unreliable data, the Civil War comes as a severe shock to the metabolism. It provides an unexpectedly rich feast of detailed personal memoirs, a cornucopia of fine regimental histories, and a solid dessert of circumstantial after-action reports. The student may gorge himself on these delicacies until he can move no more, yet still find that he has barely scraped the surface of what is on offer. In this respect, at least, the Civil War can indeed be seen as the first modern war. The spread of education and the desire to record personal experiences on paper is here exponentially greater than anything seen in Napoleonic times, even among Wellington's endlessly scribbling light infantry. This wealth of first-hand evidence, furthermore, has been lovingly preserved and sifted by succeeding generations in a manner that puts modern Napoleonic studies to shame. Civil War history remains a living subject today, whereas Napoleonic analysis was all but killed off in 1914.[6] Indeed, this qualitative difference between the way the two eras have been studied may perhaps have helped to conceal their essential underlying unity. Whereas the Napoleonic campaigns have been subconsciously relegated to a distant past, those of the Civil War are still being discovered in all their freshness from primary sources — lending them an air of modernity that may be subtly misleading.

In preparing this book I have been overwhelmed, and indeed humbled, by the generous scale of assistance and inspiration I have received from all sides. My friends in the USA have showered me with sound advice, solid interpretations and pertinent photocopies, while my friends in Britain have kept me going with their moral support mixed with searching questions. I would like to take this opportunity to thank everyone who has been involved in this process, especially the three prime movers who unwittingly pushed me into the Civil War in the first place and to whom this book is respectfully dedicated: Professor Gunther E. Rothenberg of Purdue University, Steve Fratt of South Lyons, Kansas, and Joe Park of Austin, Texas.

My first, unforgettable, introduction to the Civil War was administered by Stephen Connolly on Merseyside some twenty years ago, and he has also recently sent me valuable pointers from his new home in South Australia. Other early influences include Barrie Jones of Liverpool, Professor John Shy of the University of Michigan, Tyne Tees Television's 'Battleground' series and the late Antony Brett James at Sandhurst. Dr

Richard Cawardine of Sheffield University was particularly helpful in 1973 when he persuaded me to organise my thoughts on the Civil War, although I fear that at the time he was sadly disappointed in the result. I trust that this second shot at the target will prove to be a little more satisfactory.

I have received invaluable ideas and bibliographical advice from Richard Munday and de Witt Bailey of the Historical Breechloading Smallarms Association in London. The former's presentation of the case against long-range marksmanship (in a lecture at the Imperial War Museum, London, in July 1985) was outstanding, while the latter's technical explanations were as inspiring as they were revealing. Also in London, Ian Greenwood and Andrew Grainger were both of great material assistance over a long period. In Camberley I received help from Graeme Rimer of the Tower Armouries, David Chandler and Christopher Duffy of Sandhurst, John Drewienkiewicz and the members of the Wilkinson Society at the Staff College and – of paramount importance – the Sandhurst and Staff College library staffs. Without the unstinting labours of Mrs Kathy Clarke, in particular, the book would have been impossible to research. Also in Camberley, Anne Nason made an excellent job of the typing.

From elsewhere in Britain I was given useful information by Phil Barker of Wargames Research Group (WRG) as well as all the other members of the third Knuston Hall military history conference, held in October 1985.[7] Peter Dennis read my text and provided informed encouragement at key points.

From America I have been helped particularly by John Koontz in Colorado, Bill Haggert and Ned Zuparko in California. During my tour of some of the Civil War battlefields, I was given exceptional hospitality and assistance in the Washington DC area by David Kirkpatrick, Wally Simon, Jim Arnold, and Charles McCarthy, and in Georgia by Howard Whitehouse. No less helpful were Ed Bearss, Ed Raus and Dennis Kelly — all of the National Park Service. My thanks are also due to Ken Sirett, Jim Getz, Col. Joseph B. Mitchell, Tam Melia, Kim Bernard Holien and, in Venezuela, Walter Compton.

The book was originally published in Britain in 1987. For this new, American edition I have taken account of the perceptive suggestions made by Charles Grench and his readers for Yale University Press, and have made a number of corrections. After due consideration I have revised the work lightly to remove some of the obvious mistakes, although this does not affect the general thrust of the argument.

I am extremely grateful to all the above for all their help. They cannot, of course, be held responsible for any of my factual errors or misguided interpretations, which are entirely my own.

Equally I wish to make it clear that the book contains only my own opinions and not those of the British government or Ministry of Defence.

Finally, much nearer to home, I am grateful to my father for reconnoitering the Atlanta battlefield Cyclorama; to my wife Geneviève for her unflappability upon the discovery of a newly converted Civil War buff in her family, and to my son Robert for his innovative renditions of 'Dixie' three years before he had learned to whistle.

PADDY GRIFFITH
A 'Chancellor House' in Camberley,
Summer 1986

INTRODUCTION

The Alleged Origins of Modern Battle

... he could not put a whole faith in veterans' tales, for recruits were their prey. They talked much of smoke, fire and blood, but he could not tell how much might be lies.

The Red Badge of Courage, p. 11

The Civil War means many things to many people, but for the majority it probably stands for 'modernity' most of all. It represents the triumph of a dynamic industrial power over an outdated and quaintly chivalric order of society. It is seen (*almost* accurately) as the first war in which railroads and steamships played a significant part. It was the first war in which a recognisably modern press was exploited by populist political parties to mould the strategies of governments. It was the first war of ironclad warships and submarine torpedoes. It was the first modern war in so many different ways that we find this description almost impossible to avoid.

When we turn to the land battles of the Civil War we discover yet another element of modernity – perhaps the most telling of all – in the newly indecisive nature of combat. The apparent failure of attacking armies to defeat defending armies, even when there was a heavy numerical advantage in favour of the attack, has often been hailed as the dividing line between the warfare of the past and that of the present, the moment at which Napoleonic conditions ceased to apply and First World War conditions took over.

In this book we will examine the military and technical reasons why the Civil War battles seemed so indecisive, in the hope of casting some light on the question of just how 'modern' the war really was. Before we move on to the detailed discussion of tactics, however, let us first consider the way in which these battles are normally viewed and the sort of public image they enjoy.

IMPERSONALITY, HORROR AND DEATH

The popular image of the Civil War shows a ragged line of blue-clad infantry, battle flags waving overhead, racing across an open field towards an improvised entrenchment which is winking and flashing with the rifle fire of the defenders lurking behind it. Some attackers fall. The others try to press forward, but the unforgiving hail of lead is too much for them and the whole line hesitates. More men are hit and it is only with difficulty that the colours are kept aloft. Gradually the attackers' will to advance is crushed by the defenders' superior fire power. The attack loses conviction and melts away, leaving a corpse-strewn field and a sense of either anger or sorrow for the futility of the whole exercise. Surely the officer who ordered this butchery should have realised that the breasts of his soldiers were no match for the onward march of military technology in the age of science!

This is the scenario which we have received from generations of history books, novels and films. It is doubtless heavily charged with emotion and mythology, as all such vivid generalisations tend to be; yet it is nonetheless the picture of this war to which our thoughts most naturally return. It is a vision which owes something to our picture of Napoleonic combat, with its flags, the ardour of its charge and the shoulder-to-shoulder solidarity in its advancing line. It is a colourful image, too, since the sun is assumed to be shining, the bayonets glittering brightly, and the smart blue uniforms contrasting proudly with the golden stubble underfoot. Yet in important ways it is also very different from the brilliant Napoleonic stereotype. It contains some darker shades – and even some threatening portents of doom. It is stained with incompetence and futility, horror and death. Its palette of hues is incomplete without the dull brown mud of the defenders' earthwork, the steel-coloured smoke of their remorseless fire – and the bloody red hands of the attackers' commander as he rides uncomprehendingly away from the carnage he has produced.

A central element in this picture is its ominous impersonality, since we do not actually see the enemy who is doing his shooting from a place of safety. He is concealed behind his works, if not actively camouflaged: he sees but is not seen. Consider the following passage from a first-hand description of battle published in 1864 by John William De Forest, a Union infantry captain who was also a novelist:

We were just entering a large open field, dotted by a few trees and thornbushes, with a swampy forest on the right and the levee of the bayou on the left, when the Rebels gave us their musketry. It was not a volley but a file fire; it was a continuous rattle like that which a boy makes in running a stick along a picket fence, only vastly louder; and meantime the sharp *whit whit* of bullets chippered close to our ears. In the field before us puffs of dust jumped up here and there; on the other side of it a long roll of blue smoke curled upward; to the right of that the grey smoke of artillery rose in a thin cloud; but no other sign of an enemy was visible.[1]

Another novelist, although not one who had personally seen action, returned to make a similar point in 1894. During the attack in Stephen

Crane's *Red Badge of Courage* there is an even greater sense of dis-orientation:

The flaming opposition in their front grew with their advance until it seemed that all forward ways were barred by the thin leaping tongues, and off to the right an ominous demonstration could sometimes be dimly discerned. The smoke lately generated was in confusing clouds that made it difficult for the regiment to proceed with intelligence. As he passed through each curling mass the youth wondered what would confront him on the farther side.[2]

The 'thin leaping tongues' in this passage are the tongues of flame leaping from the muzzles of the hidden enemy's rifles as they deal death only too effectively on all sides. We thus have a numbed, sightless and vulnerable regiment in the open confronting an unidentified and invulnerable murder machine. This is scarcely a 'sporting' way to fight a battle – if 'fighting' is an appropriate word to use at all in the context – since the attacker is deprived of much of his free will by the hostile environment into which he is plunged. His chances of defeating the enemy appear remote,[3] and he takes on the quality more of a victim than of a warrior. He becomes a pawn to be sacrificed in a game which he does not understand and which we suspect his commanding officer does not understand either.

The impression is created that this kind of warfare must have seemed strange indeed to a Civil War generation raised on stories of Napoleon's lightning victories and brilliant assaults. The temptation must have been strong to suppose either that 'someone had blundered' when these masterly manoeuvres failed to reappear in the 1860s, or at least that something had changed in the physical process of combat. Either way, the newly empty battlefields could be portrayed as evidence that generals no longer controlled the outcome of their tactics, and hence that there had been something of a revolution in warfare since Waterloo.[4]

Whatever the strength of feeling on this point may have been at the time of the war itself, it is certainly very difficult to find any historians working today who are prepared to disagree. The accepted reasoning has it that, because Civil War battles were apparently more indecisive and impersonal than Napoleonic ones, something important must undoubtedly have changed in the nature of tactics. Despite the fact that Civil War battles were less lethal than their Napoleonic models – and fought in very different geographical and strategic circumstances – the corrosive idea remains widespread that the Civil War somehow represented nothing less than the origin of modern battle itself.

We do not have to look far to find historians who claim that 'Tactics lagged behind technology in the 1860s', or that 'Civil War assault forma-tions were obsolete in comparison to the fire against which they were launched'.[5] A new imbalance between attack and defence is thus deemed to have produced the particularly indecisive series of battles for which the war is notorious, and from which arose both the unusual length of the conflict and the widespread suspicion of incompetence adhering to its generals. They might have been clever enough to manoeuvre their forces into contact with the enemy, it is argued; but once there, they usually failed to finish him off. Worse still, if they tried to press their attacks too strenuously they risked debilitating casualities to their own men without any worthwhile gain against the opposition.

An interesting recent variation on this theme, in a study by Grady McWhiney and Perry D. Jamieson, would have it that the propensity to launch ill-considered and costly assaults was a specifically Confederate characteristic: 'The South simply bled itself to death in the first three years of the war by taking the tactical offensive in nearly seventy per cent of the major actions.'[6] The authors go on to attribute this in part to the Celtic origins of southern society. The wild, yelling charges at Telamon, 225 BC, and Culloden, AD 1746, are seen as racial ancestors of Pickett's charge, while the Rebel yell itself is given a venerable pedigree which few of its practitioners can have suspected at the time. Admittedly this 'Celtic' interpretation of Civil War tactics has found many detractors, but it remains true that the Confederates have always enjoyed a reputation as particularly tough and aggressive fighters, and can claim considerable support from the evidence – especially in the Eastern theatre – to back up that reputation. Most authorities do also agree that something had indeed gone wrong with the art of attack by the time we reach the 1860s, and that battles had become 'at once much longer in time and less decisive in outcome. There were to be no more Waterloos.'[7]

All this means that in the popular view the Civil War was moulded by a failure specifically of tactics, as opposed to strategy or administration. By their inability to deliver a climacteric 'Napoleonic' battle, it was the tactics which made it inevitable that the war would be decided elsewhere than on the battlefield:

Tactically, Civil War armies indeed were able to hurt each other – sometimes a great deal – but mutually and indecisively. Poorly chosen tactics rendered higher casualty lists on both sides. *The war was decided by the formulation and execution of superior strategy, psychological damage, and scale-tipping impact on logistics.*[8]

What these 'strategic, psychological and logistic' measures implied was nothing more or less than the birth of total war, or, in less specialist language, a war which spilled over into attacks on civilians and non-combatants. The destruction of public property merged only too easily into the imposition of punitive measures upon enemy sympathisers – and then into attacks on *suspected* sympathisers. This was a different aspect of the horror of the Civil War from the impersonality of the battlefield, but it was one which flowed directly from it. Because the armies in Virginia could not reach a quick decision, that state was systematically devastated

18

for four weary years. Hogs, chickens and fence rails became prized curiosities of exceptional rarity. Generals took to boasting that if a crow wished to fly across those lands it would have to take its own rations along with it.[9] And meanwhile in Tennessee and Georgia the same bleak process was working with even greater effect. Cities were destroyed and tens of thousands of families made destitute. Gaping wounds were slashed into the nation's cherished self-image as a fraternal, benevolent and tolerant association of free men. Nor was there even the avenging compensation of a similar loss inflicted upon some other – foreign – nation.

The end of the blood-letting did admittedly produce a certain catharsis and reconciliation, and strenuous efforts were made in the following decades to gloss over the magnitude of the disaster,[10] but ultimately these attempts have failed. Even though the United States of America today remain indisputedly united, even though most of the opposing commanders shook hands chivalrously and made it up with each other the moment the firing ceased,[11] there is still an underlying feeling that the whole affair was in reality an unjust and grisly mess. The cold-blooded destructions of Atlanta, Columbia, Richmond and Fredricksburg are no more easily forgotten and forgiven than the casual abuse of the prisoners of war or the fratricidal activities of bushwhackers and jayhawks in the frontier badlands.

Yet in the end it is still the horrors of the battlefields that we remember more vividly than the attacks upon civilians or prisoners. It is the plundered bodies of the piled dead that haunt us from the Bloody Lane or the Devil's Den – expressive place-names indeed![12] Merely the bare statistics of the casualties still possess a power to shock. Thus Antietam is remembered as America's bloodiest day, with a total of 22,000 killed and wounded on the two sides within the space of less than twelve hours. At Second Manassas the two sides lost some 24,000 between them in the course of a two-day fight; at Chancellorsville the figure was approximately 30,000 in four days; at Gettysburg around 43,000 in three days. In the first ten days of his ill-starred command around Atlanta General J.B. Hood suffered 19,000 Confederate casualties in a series of what are now all-but-forgotten skirmishes; while in his first forty days of campaigning with the Army of the Potomac, from the Wilderness to Cold Harbor, the Union General U.S. Grant lost no less than 55,000 of his troops in combat, almost double the total number of the Confederates who had earlier opposed him at Vicksburg.

These were not scales of loss which would have surprised any European soldier of the period, since at Solferino in 1859 the combined casualties of the three armies engaged had amounted to some 40,000 within twelve hours, and at Borodino in 1812 it had been nearer 75,000 within the same interval. In the biggest Napoleonic battle, at Leipzig in 1813, the butcher's bill had been around 127,000 casualties in the space of four days.[13] This sort of statistic can help us to put the American Civil War into perspective (just as it can help us to put Verdun and the first day of the Somme into perspective): but it somehow fails to erase the accepted folk memory that, in a mysterious but profound sense, the Civil War was actually the very worst of all the nineteenth-century wars.

We should remember that both mass literacy and the popular press were expanding rapidly in the later nineteenth century, and with them came a new vocabulary – a new vividness – in the reporting of warfare. Battle descriptions took on a new meaning for the common man, and we should scarcely be surprised to find that this included a heightened stress upon the horrors and dangers of the battlefield. Even if the battlefield had not in fact become more lethal and impersonal, it would probably still have been described as if it had. Some scholars, indeed, have noted that much of the imagery and rhetoric of the First World War had already been assembled some decades before 1914, as a part of the general literary evolution of late Victorian society.[14] In this sense it is quite reasonable to suppose that writers like Crane, who were searching for images with which to describe the Civil War, were also helping to write the script for our subsequent perceptions of the First World War.

EFFETE EUROPEANS AND PERVERTED PROFESSIONALS

It is certainly to the First World War that most people turn for a parallel to the horrors of the Civil War, rather than to the 'cabinet' wars of the European past. We repeatedly encounter such statements as 'The war became one of attrition, typical of what took place later in World War I', or 'Grant's decision to seek confrontation in a frontal fight made the campaign from the Wilderness to Petersburg a dress rehearsal for World War I.'[15] Equally, we are often assured that 'No war of the Eighteenth century had been like this',[16] that its battles produced 'record bloodbaths',[17] and that:

The American Civil War was the first major conflict of the industrial era; the first to see extensive use of railroads, telegraphic communications, fast-firing weapons, armored warships and machine-made uniforms, boots and horseshoes. The fratricidal clash of 1861–5 witnesses the shift from the war of movement of Napoleon and Winfield Scott to the war of attrition of Grant, Sherman and the Western front.[18]

According to this interpretation the Americans were wrestling with the problems of modern warfare over half a century before Europeans were forced to confront them:[19]

'In understanding the power of the defense, American officers in this war proved far more perceptive than later high commands in Europe. In spite of the development of the machine gun and quick-firing artillery with improved shrapnel, European soldiers a half century later in World War I would have great difficulty in understanding the by-then quite obvious and almost impregnable power of the defense.[20]

Or, again, even more forthrightly:

Since our civil war the thick headed English generals, failing to profit by our experience, decimated their armies in South Africa assaulting the Boers in fortified positions until Lord Roberts took command, flanked their fortified positions and captured Pretoria.[21]

The 'thick headed English generals' are not the only victims of this sort of rhetoric, although they are possibly in the bull's-eye of the target. The French and Germans also come in for their share of denunciation, especially since France (in the person of General Bazaine) was actively involved in Mexico at the time of the Civil War, and Germany attempted to become so in 1917. Such breaches of the Monroe Doctrine have been taken as pretty clear challenges to the American military establishment, and unmistakable ones when we add that no less an officer than the Prussian von Moltke the Elder was prepared (albeit apocryphally, as it transpires) to dismiss the Civil War as 'two armed mobs chasing each other around the country, from which nothing could be learned'.[22] The implication of this squib is obviously that the undisciplined national character of the American soldier is such as to unfit him for any sort of comparison with the military bearing, the solidity or firmness of step of the Potsdam grenadier. It is a view which has been widely repudiated, and we often read that — at least by the middle of the war — the American armies 'were fully equal in quality to the forces fielded by the great military nations of Europe',[23] or that 'the art of war had reached a higher stage of development in the United States than it reached in Europe by 1866, and in some respects, higher than it attained in 1870'.[24]

This brand of analysis does not linger long over the tactical naivety which was unfortunately so prevalent in Pershing's army as it finally entered combat in France in 1918, nor does it face up very clearly to the heavy questioning of officer competence which emerged from the Civil War itself. Instead, it prefers to refer back to one of the oldest American stereotypes of all – the image of the European officer as aristocratic, brainless, over-concerned with 'good form' and under-concerned with finding realistic solutions to the military problems which he faced.

Nor is it only the real Europeans who have been vilified in the Civil War debate, since American soldiers and institutions which tried to imitate European models have also been tarred with the same brush. It is easy to recall that the drill books used in the Civil War were almost verbatim translations of French originals, leading to a set of European tactics being artificially imposed upon the armies of a completely non-European continent. Equally, the higher theories of war allegedly applied by many American commanders in the 1860s were almost entirely made in France.

21

It is Bonaparte the renegade Corsican and Jomini the renegade Swiss who are generally supposed to have written most of the books which shaped the battles. Such all-American figures as Dennis Hart Mahan take at least third place when set beside these colossi of the military art. There is no avoiding the fact that American military institutions before the Civil War were moulded most profoundly by the military theories of the French, and it is therefore the French who take a major part of the blame for the military disasters.

At a more personal level, it is almost as easy to identify particular individuals who were deliberately trying to ape the European code of officer behaviour. Extroverts like McClellan and braggarts like Pope may readily be criticised in these terms, but behind such criticism there also lies a wider suspicion of the whole system of officer education in peacetime. West Point has especially been accused of fostering an aristocratic, un-American military caste, and it was noted in loyal Northern circles that many of its products shamefully betrayed their oath of allegiance by fighting for the Confederacy. Among Southerners there was an equal and opposite feeling that the Military Academy had somehow exerted an alien, Federal influence upon many of their most favoured sons.[25] It was deemed to have blunted some of the common sense and robustness of the Southern fighting man, and given an unfair title of advancement to many mediocre officers:

[General Lee] failed to realize that while a military school is excellent for the training of drill masters, who are most necessary, it teaches little of military science in comparison with the hard experience of a single campaign . . . it was often difficult to get past the incompetents, who went in as drill masters and were then pushed up by the ability of their men, who were of a class few armies have ever seen, to become commanders of large operations.[26]

Or again:

General Lee never went outside . . . the regular grades to find officers, who might have been very Samsons to help him multiply his scant resources. He never discovered or encouraged a Nathan B. Forrest, and many a man went to his death trying to win against the incompetency of leaders who should have been brushed out of the way when they failed.[27]

It is paradoxical that General Lee should be criticised on these grounds, since it is precisely the closeness of his circle and the excellence of his military pedigree that are traditionally counted among his highest qualities.[28] As the epitome of the Virginia gentleman he stands unrivalled, while in his brilliant battlefield manoeuvres we can surely see a triumphant vindication of his West Point education. If we look more deeply into the accusation, however, we may still glimpse the particular truth which the critics of West Point were trying to illuminate. The acid fact is that Lee lost the war. His Southern chivalry and his virtuoso tactical footwork were ultimately overwhelmed by novel forces which Southern society – no less than the old West Point – was ill equipped to meet. The Northern blockade, the Northern industrial mobilisation and the clear-sighted Northern policy of attrition – these were the iron imperatives which finally brought Lee to Appomattox. For many commentators, therefore, the war has become a parable in the futility of such romantic notions as 'nobility' or 'officer-like behaviour'. It was those officers who broke away from their professional code of chivalry who finally came out on top, not those like Lee who remained loyal to it. This *was* the first modern war, after all.

The nub of this critique seems to be the belief that West Point was more concerned with social status and outward appearances than with hard objective realities. If its highest priorities were rote-learning and 'style', the Academy could scarcely hope to equip its products with a robust perception of the rapidly changing military universe. If the lessons taught in West Point classrooms were those of the Napoleonic or Mexican wars, then West Point officers might be expected to commit disastrously anachronistic errors when they arrived on the battlefields of the 1860s. According to many commentators of the Civil War, indeed, this is exactly what did happen. There is a profound impression that many officers were completely divorced from tactical reality; they operated entirely in accordance with some remote theoretical system. One recent author has even gone so far as to state that 'Losses were so staggering because officers on both sides fought by the books, and the books were wrong.'[29]

On the other side of the same coin we can find an equal and opposite panegyric to the American ideal of the plain man, the practical, down-to-earth frontiersman. According to this interpretation it was the soldiers – the simple farmers and artisans who filled the ranks – who could see deeper into the true situation than their distracted and etiolated commanders. Even better were such untutored but innovative leaders as Nathan B. Forrest, the homespun hero of the Western theatre. Best of all, in fact, were those few West Pointers who consciously and deliberately set their book learning to one side – unpretentious men like Sherman and Grant – in order to knuckle down and win the war as it was really fought. These men tended to owe their promotion to their activities in the Western theatre, well away from the eye of the Washington politicians. They had often spent long years before the war on the unrewarding Californian frontier, and they would typically have little time for the ceremony and pomp of peacetime professional soldiering. One Union enlisted man summed up their style in the following words:

Those of our Corps who served two years in the Army of the Potomac before they came west noticed the great difference between the two armies, especially in the dress and conduct of the general officers, all the way from the Brigadiers to the General in Command. In the east our Generals, as a rule, made a real military show with brilliantly dressed staffs. They followed after the General who usually was in full dress with sash and sword and all the buttons allowed his rank. The staff officers were followed by an escort of Cavalry varying in number according to the officer they followed in the parade. In the Western Army I never saw such a gaudy show; there they seemed to avoid show of any kind.[30]

If the style of fighting which these 'Westerners' embraced was actually less elegant and more destructive than that of their effete eastern colleagues, then that was the tax which common sense had to pay to changing circumstances. Americans could at least congratulate themselves that their own national virtues had triumphed over imported foreign influences: the challenge of the new warfare had been successfully overcome.[31]

All this has a ring of familiarity for those who have read one of the most remarkable of all the books written in the 1860s – Leo Tolstoy's immortal *War and Peace*. It is true that Tolstoy does not concern himself with America or her problems, yet in his pages we find that in the Russia of 1812 it is the earthy common sense and homespun philosophy of the patriotic peasant that triumphs over the artificialities and shallownesses of the western European military advisor and courtier. Napoleon's new style of warfare shatters the old pedantic drills of the German experts, but it is met and defeated in turn by the united willpower of the Russian masses. Although Tolstoy places the appearance of 'modern warfare' some fifty years before the bombardment of Fort Sumter, he is clearly reflecting the same impatience with the aristocratic 'lace wars' of the eighteenth century as the Americans experienced in their own conflict. In both cases the culprit is located in a courtly west European system of values which has outlived its usefulness. In neither case is the old European order seen to be truly relevant to the new forces straining to emerge around the fringes of the Old World.

GROUNDSHEETS, GADGETS AND GUNS

The feeling that the Civil War was the first modern war may certainly be attributed in part to the emotions which come naturally to a thriving young country anxious to kick off the influence of its former colonial power. It is often a feeling born of national pride and patriotic rhetoric. Yet beneath it all there remains a technological strand in the argument – a strand which has more to do with the industrial revolution of the 1850s and 1860s than with the political, anticolonial revolution of 1776.

In the mid nineteenth century armies were for the first time coming into contact with a wide range of new gadgets and mechanical wonders which – it was hoped – would suddenly take the labour and danger out of campaigning. The French in Algeria had invented the cholera belt and the portable one-man shelter tent (a notable advance over the wagon-carried sixteen-man Sibley). The rubber blanket-cum-groundsheet was another enormous step forward for the bivouacking soldier,[32] and it was followed by the occasional canned peach, the intermittent desiccated vegetable and even the very rare can of condensed milk. Admittedly not all of these inventions were well received by the troops, who had to camp rough and travel light through all 365 days in the year; but it could scarcely be denied that they were precursors of a technological change in the whole environment of the camp and the march such as had not been dreamed of by Napoleon's *grognards*.

Even marching itself was in many instances made unnecessary, since the railroad and the paddle-steamer could now be used to spirit troops from one end of a theatre of war to the other in a matter of days, whereas previously it had taken weeks. Aficionados might notice that there were severe risks attached to 'riding the cars' – particularly in the Confederacy – and that trains regularly jumped off the tracks.[33] But, when all things were considered, the railroads beat walking by such a handsome margin that no one would have wished to do without them.

Along the railroads ran the electric telegraph wires, and at the terminals of the telegraph sat dedicated professional signallers who had accepted a measure of military supervision in return for exemption from the draft. By 1865 the US Military Telegraph represented an astonishingly wide-ranging and comprehensive communications service, feeding central government with strategic intelligence hourly and in great detail.[34] Less spectacular, although scarcely less important from the tactical point of view, were the great strides which had been made in the portable field semaphore – the 'wig wag' – with lights by night and flags by day. This perfected system of tactical communications made it possible to send short messages across a battlefield within half an hour or so, even without such exotica as balloons and secret ciphers, each of which also seemed to represent a new (if erratic) dimension in signalling technique.

Still more suggestive of the First World War were some of the 'secret weapons' which were being tested in the 1860s. Quite apart from the naval innovations – the ironclads, submarines and torpedoes – we find several different types of machine guns, hand grenades, telescopic-sighted sniper rifles, explosive bullets, gas shells and trench periscopes, and at least one railroad-mounted siege mortar.[35] In 1862 in the Yorktown peninsula the pyrotechnical Rains brothers planted some pressure-sensitive ground mines (also called torpedoes) in the path of the advancing Union army.[36] At Fort Sanders, Knoxville, in late 1863 the Federals achieved some success with defensive wire entanglements,[37] a trick which they repeated six months later at Bermuda Hundred.[38] None of these gadgets really had any significant effect upon the course of the Civil War – nor, in fact, did any of them catch on seriously until at least a generation after 1865. Nonetheless they did each put down notable markers for the future and have caught the imagination of students on the lookout for the first modern war.

Less suggestive of the trenchlock of 1914–18, but of enormously greater importance to the Civil War itself, were two other new families of weapon which did see more or less widespread service between 1861 and 1865. The first consisted of rifled muzzle-loading cannons and muskets. These had gradually been replacing the traditional smoothbores since the 1840s, until by 1865 almost all infantrymen were armed with either the ·58 Springfield or the ·577 Enfield rifle musket. (The artillery, however, still retained a very large proportion of smoothbore cannon.) Secondly, there was a bewildering variety of new breechloading carbines by the end of the war. Only a few of the infantry were so equipped in 1865, but in that year the US Board of Ordnance was finally converted entirely to breechloaders. The superiority of this arm over the rifle musket was acknowledged to be as great as that of the rifle musket over the smoothbore musket and the only point remaining to be decided, it seemed, was whether the infantry's breechloaders should be repeating or single-shot. In 1865 the decision went in favour of the single-shot weapon, although this outcome has subsequently been denounced with some vigour by writers on firearms. It has also been pointed out that 'the first modern war' would actually have been considerably *more* modern if only the Union army had embraced the breechloader for its infantry in 1861 rather than waiting until the end of hostilities. Some even go so far as to suggest that the war could have been won quickly and cheaply if this had happened, since it would have given the Union a clear technical advantage which the Confederacy could not have matched.[39]

A cynic might be tempted to suggest that much of the fuss about new weapons in the Civil War has been unduly inflated by a rare combination of hindsight, blind faith in gadgetry, and unrestrained commercial hype. The oft-quoted testimonials to the effectiveness of breechloaders certainly owe much of their fame to this last influence, since wartime production of such weapons was entirely in the hands of private firms such as Colt, Remington, Winchester and Spencer. These unashamedly entrepreneurial businesses easily lapsed into many forms of cutthroat competition with each other in the search for private, state and federal contracts. Hawkers pestered raw recruits with ingenious new collapsible pistols for their 'secondary armament', while Spencer twice visited President Lincoln to give him a free shot with his patented repeating carbine. Shady wheeling and dealing for contracts was commonplace, and so was the collection of testimonials from satisfied customers. We thus have some neatly documented accounts of the wonders of the various breechloaders, and their battlefield efficacy, in a way which is not quite true of the more mundane equipment manufactured in government armouries. To a certain degree this imbalance of propaganda has distorted and inflated our perception of the more advanced weapons.[40]

Introduction

There is also a distinctly patriotic flavour to the widespread denigration of the alternative to modern firearms, namely the *arme blanche*. Just as the British in the Revolutionary War were condemned for their reliance on bayonets and the colonists praised for their alleged use of the Kentucky long rifle,[41] so the American writers on the Civil War have formed a solid front against the cold steel. We read that 'any saber was culturally alien to most Americans'[42] (although not, apparently, to their West-Point-trained officers)[43] while in many cavalry units it almost became a mark of 'frontiersman' status to do without it.[44] The same, to an even greater extent, can also be said of the bayonet. Its pretensions in the war are today almost universally mocked, and we repeatedly find quotations from the Union medical statistics of 1864 which show how utterly negligible were the number of wounds which the bayonet caused.[45] As for such implements as pikes and bowie knives, they enjoyed a certain vogue in the South at the start of the war when there were no other weapons available,[46] but the general verdict on them can be gathered from the following terse extract from the *Richmond Daily Examiner*, 13 August 1861: 'The Tiger Rifles, Wheat's battalion, charged with Bowie knives at Manassas. But 26 out of 83 rank and file survive'.[47] If the emphasis in the literature is all upon firearms – and the more modern the design the better – that is perhaps perfectly understandable. It was guns that did the actual killing and the killing, presumably, dictated the shape of the battles. Time and again, therefore, we find the battles interpreted in terms of a 'revolution in firepower' or of a 'rifle revolution' which is assumed to have taken place around the early 1860s.[48] This in turn is assumed to have precipitated the tactical stalemate which did so much to give the war its 'first modern' profile. The rifle, in other words, is conventionally seen as the key to the whole conundrum. It is the factor selected by historians as the one most likely to explain the supposed changes in the nature of combat during the Civil War.

In the present volume we will discover that these claims for the influence of technology on the Civil War battlefield are, to say the least, anachronistic and exaggerated. They have rather more to do with late twentieth-century habits of thought than with the military realities of the 1860s. As in so many other eras of military history, in fact, it transpires that human factors such as training and doctrine – or the lack of them – exerted a much greater underlying influence upon the outcome than did the precise specifications of the weaponry in use. And by the same token it also transpires that the (very human) prejudices of propagandists and historians have generally tended to keep this reality hidden.

1

The Armies
Learn to Fight

'B'jiminey, we're generalled by a lot 'a lunkheads.'
'More than one feller has said that t'day,' observed a man.
The Red Badge of Courage, p. 76

It does seem to be the case that tactical attacks in the Civil War attained their local objectives rather less frequently than those of Napoleonic times,[1] although that does not necessarily mean that they were less useful in determining the final result of battles or campaigns. Lee's offensives at Chancellorsville made little tactical sense, despite some splendid local successes, yet they produced a remarkable strategic victory. Conversely, quite a large number of promising attacks were called off before they had achieved victories which were well within their grasp, while still others were thrown away by strategic failures after a positive tactical result had been secured. If the offensive seemed to be rather a disappointing instrument in the hands of Civil War commanders, this was not always the fault of the offensive itself.

It is worth remembering that much the same can also be said of the defensive, since the Civil War commanders often could not or would not convert a successful defence into a crushing victory. Lee at Fredericksburg and Meade at Gettysburg both repelled the enemy's assaults as completely as had Wellington at Waterloo, but in neither case did they take up what Clausewitz called 'the flashing sword of the counter-offensive'. They enjoyed outstanding advantages over their opponents, but they did not win outstandingly decisive victories. The decisiveness or indecisiveness of Civil War tactics should not therefore be confused with the properly separate question of whether attack or defence was more powerful.

In order to help us clarify these matters a little more fully, we will now briefly review the course of the war and the various ways in which battles were fought. In particular we will examine an important 'human factor' in the process – the way in which the senior commanders learned to apply a specific personal style in the way they tackled these problems. We do not have to look very far to find wide differences in approach between the generals. Some were indelibly cautious, even before they encountered the shock of battle. Others remained wedded to the attack despite all experience and many wounds. Some were 'mud-diggers' from start to finish, while others were confirmed believers in the bayonet. Others yet again developed personal variations and changed them in the light of changing battlefield conditions. The spectrum of individual preconceptions is therefore almost infinite, and we should beware of making too sweeping generalisations about the 'typical' Civil War commander and his tactical views. There was no such person as the typical commander, any more than there was such a thing as the typical Civil War army.

THE FIRST CLASHES

History[2] tells us that the war started on 12 April 1861, when the states which had seceded from the Union finally resorted to force in their attempt to remove Union garrisons from their soil. Since the peacetime strength of the US Army was a paltry 16,000 – and few militias had yet entered the field – the fighting was necessarily limited in scale at first. Both sides set to the task of improvising armies from almost nothing, and the wonder is that they could between them assemble some 85,000 men for the First Manassas campaign, as early as mid-July. That represents a speed of mobilisation comparable to some of Napoleon's most doomed and desperate measures in the crisis of 1814, although in that case, of course, there was a solid *cadre* of veterans with two decades of combat experience to add strength to the new units.[3] In 1861, by contrast, the veterans were so heavily outnumbered by the newcomers that their influence may be regarded as negligible, while the main combat experience came from a small war in Mexico fifteen years before. When the armies met on Bull Run on 21 July, therefore, it was inevitable that they should have been amateurish, clumsy and – above all – fragile.

The most important point to notice about the First Battle of Manassas, however, is probably not that the soldiers of both sides were unsure of themselves and vulnerable to wild rumours and panic, but that the Federals happened to be the ones who panicked first. This single event – almost a pure chance – had incalculable repercussions throughout the remainder of the war. Its importance cannot be overstated, not merely because it robbed the Union of a quick victory but – far more significantly – because it set the tone for everything that was to follow. Having lost the first major contest the Federals formed a low opinion of themselves and a correspondingly high opinion of the Confederates.[4] Still more to the point, perhaps, was the fact that the Confederates came to precisely the same conclusion.

The outcome of the first battle of a war is almost always a matter of paramount importance to the balance of morale between the two sides. Whether it is the British imagining the ten-foot-tall Japanese soldier after the fall of Singapore in 1942, or the Italians imagining the ten-foot-tall Tommy after a few skirmishes on the Egyptian border in December 1940, the effect on subsequent fighting quality is enormous. Even if he enjoys overwhelming material advantages, the loser of the first clash will normally find it very difficult indeed to rebuild the necessary self-confidence in his men. Conversely, the winner of the first clash will imagine himself invincible, which goes a long way to making him so.

This truism was well understood in 1861 by a certain John B. Jones, a clerk working in Jefferson Davis's War Department. On 18 June, a month before First Manassas, he entered this prophetic note in his diary: 'We *must* win the first battle at all hazards, and at any cost; and, after that, – how long after? – we must win the last!'[5]

Once the Confederacy had indeed won the first battle it found that it possessed a magnificent 'force multiplier' in almost all its meetings with the enemy, at least in the Eastern theatre, where the close proximity of the two capital cities, Richmond and Washington, tended to make the fighting more concentrated than elsewhere. A typical tale of Southern morale in August 1861 is recounted by the Confederate private, Sam Watkins, when his officer used a seven-shot repeating rifle (presumably a Spencer) to shoot an exceptionally large number of enemy soldiers in a skirmish:

I have forgotten the number that he did kill, but if I am not mistaken it was either twenty or twenty-one, for I remember the incident was in almost every Southern paper at that time, and the general comments were that one Southern man was equal to twenty Yankees.[6]

Returning to First Manassas, we find that both sides had originally intended to attack, but the Federals used better staff work and landed their blow first. Things then started to go wrong for them, and it was the defending Confederates who eventually won the day. The latter managed to stabilise a line, 'like a stone wall', in fairly strong terrain which was a mixture of woods, small streams and open fields. There were very few fortifications in this battle, although the Union commander, McDowell, believed in their potential value and his men apparently had a great fear of them – as well as of some purely mythical Confederate 'masked batteries'.[7] Instead of all this, the power of the defence at Manassas actually stemmed from a mobile or active concept of operations – the concentration of reserves from unengaged sectors of the line against an overambitious attack by a flanking echelon. McDowell had clearly studied his Frederick the Great, but he failed to notice that Frederick's grand flank attacks had usually relied upon an exceptionally well drilled and experienced army. At First Manassas there was no such army available, so the attacks went in piecemeal and ineffectively. Because the Union manoeuvres were more complex and demanding than the Confederate response, they fell apart more quickly. The Confederates, on the other hand, lacked the confidence necessary to make an effective pursuit.

Apart from this single – highly revealing – battle, the remainder of 1861 was taken up with relatively small-scale skirmishing, as each side tried to stake out its claims to as much of the border territory as it could. In several cases a whole state could be dominated by no more than a couple of thousand men, simply because the opposition could field only a few hundred. Also – and ultimately of great significance at the strategic level – the Union started to apply General Winfield Scott's famous Anaconda Plan for the blockade of the South's coastline and the seizure of the Mississippi River line. The cotton trade was stifled and some important obstacles were placed in the way of gun-runners from Europe to the Confederacy.

Through all this the armies were being remorselessly built. The senior commanders kept insisting that their men could not take the field until they had received enough drill and acclimatisation to camp life to make them effective as soldiers. The losses to disease alone were horrific in this period, but so were the losses from front-line service due to such things as 'temporary' detachments for rear echelon duty, or the disbandment of units which had signed on for only three months' service. It took much time and patience for the armies to shake out and find their identities, for the dedicated men to rise to the fore and the shirkers to reveal themselves.

There was also a problem of armament on both sides, since the best weapons had been distributed so quickly that supply soon broke down under the pressure of runaway demand. By March 1862 the two sides between them had over half a million men in the field, and many more deployed on the lines of communication or still coming through the recruiting centres. This must be set against a total of some 700,000 infantry weapons available in USA before secession, of which about half a million were obsolete smoothbores and less than 36,000 were modern rifles of the optimum 0·58 calibre. Obviously the need was to find well over a million rifles within a few months, at a time when normal US production was running at only around 15,000 per year.[8]

Lincoln was anxious to push forward – 'On to Richmond!' – as soon as he possibly could. He chafed at the delays imposed upon him by the mundane business of creating an army, and his confidence in the man who was creating it began to wane. This man was George B. McClellan, an outstanding and highly educated soldier who had the art of war at his fingertips, although he did display some unfortunate qualities of character when the pressure began to tell. McClellan was popularly believed to be the nearest thing to a new Napoleon that the Civil War could produce, and this was an impression that he himself did nothing to dispel. He was loved by his troops and convincing in his military analyses, but – alas for the Union – he ultimately turned out to be more like Napoleon III than

Napoleon I. He could be as brilliant as Napoleon III on occasion, but he also suffered from the same intellectual doubts and hesitations as that officer, and his ultimate fate was the same.[9]

THE EASTERN THEATRE

McClellan finally started moving in March 1862 with his Army of the Potomac some 150,000 strong. Although outnumbering the Confederates very heavily, he nevertheless opted for a cautious, methodical approach, with the maximum use of fortification, at Yorktown and then on the wooded and swampy Chickahominy River line. He allowed himself to be overawed by exaggerated intelligence reports of the Confederate fortifications and then, several times over, by poorly co-ordinated Confederate attacks. These attacks – at Seven Pines and in the four 'Seven Days' battles – were just as ambitious as McDowell's at Manassas and technically just as unsound. The Army of Northern Virginia had not yet found the peak of its form and its commanders had still to perfect their staff work. Nevertheless, the very fact that they could attack at all was enough for McClellan. Despite his great numerical and material advantages, he packed his bags and retreated. Richmond was saved by the irrational but effective Confederate belief in the offensive.

Unlike the Confederates at First Manassas, the Northern troops in the Peninsula lacked a mobile concept of defensive warfare. Their idea was to dig into a position and fight a static battle. At Beaver Dam Creek and Malvern Hill this turned out to be a perfectly adequate tactical response to the Confederate attack, although on both occasions the army retreated soon afterwards for strategic reasons. At Gaines's Mill, however, an excellent fortified position was eventually carried by repeated frontal assaults. Because McClellan had failed to provide sufficient mobile reserves, the battle was lost. This lesson does not appear to have been properly learned by the Army of the Potomac, however, and in many of its later battles we still find a preference for a static defence based on firepower and fortification rather than on fancy footwork.[10]

On the Confederate side there were two separate lessons learned from the Peninsula battles. On the one hand, Joseph E. Johnston, who had commanded at Seven Pines, found his suspicion confirmed that it does not pay to make attacks against a superior enemy in rough terrain. Not only had his attack finally run down with some 6,000 casualties, but he himself had been badly wounded. This was a most painful way to learn caution, and it prevented him from ever becoming a fully-rounded, risk-taking commander.

On the other hand, Johnston's successor, Robert E. Lee, had learned a precisely opposite lesson from his battles. He had made four attacks in

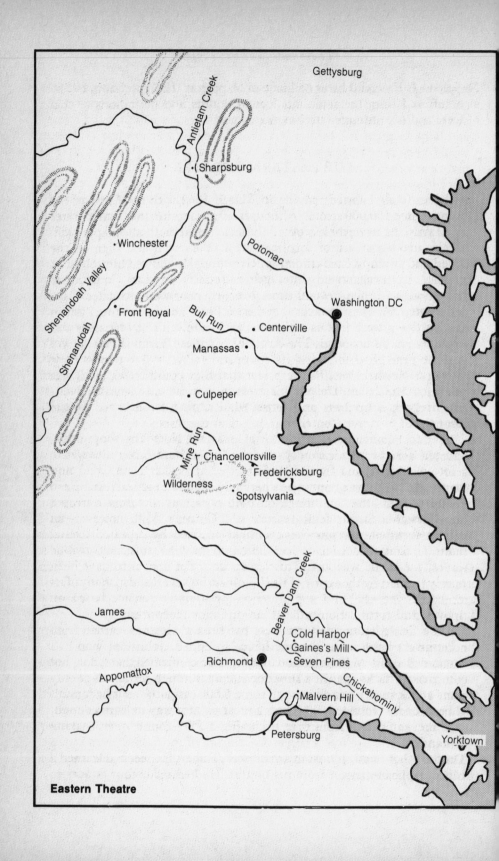

Gettysburg

Antietam Creek

Sharpsburg

Potomac

Winchester

Shenandoah Valley

Front Royal

Bull Run

Centerville

Manassas

Washington DC

Shenandoah

Culpeper

Mine Run

Chancellorsville

Wilderness

Fredericksburg

Spotsylvania

James

Beaver Dam Creek

Cold Harbor

Gaines's Mill

Richmond

Seven Pines

Appomattox

Chickahominy

Malvern Hill

Petersburg

Yorktown

Eastern Theatre

equally unpromising circumstances, but despite the loss of about 20,000 casualties he had achieved his strategic objectives. For the future he doubtless felt that better staff work and co-ordination was required, but by and large he must still have believed that the best way to whip the Union would be to keep on attacking its faint-hearted armies. We do not have to invoke any theory of Celtic racial characteristics to understand this policy, since it was merely the logical outcome of Confederate experiences up to that point. Lee had seen that his attacks defeated the enemy even when they failed to capture local objectives. Paradoxically, a tactical failure could still be a strategic success. This taught Lee that the offensive was the best way to fight battles.

Linked to the idea of the offensive were the ideas of manoeuvre and surprise. It was at this time that the celebrated partnership between Lee and General Thomas J. Jackson was formed, and Jackson above all things believed in manoeuvre and surprise. In his Valley campaign he had dazzled friend and foe alike by the rapidity of his movements and the originality of his conceptions. Admittedly, this had been a very small-scale operation and Jackson performed much less brilliantly when he arrived in the big league near Richmond. Nevertheless, his military apprenticeship had been one of unparalleled excellence, and he was soon to adapt it effectively to operations on a larger scale during the Second Manassas campaign.

In this campaign, fought during August 1862, the Confederates adopted a highly 'Napoleonic' system of march manoeuvres, with their army split into a fan of several major segments which united only on the battlefield itself. An attempt to achieve this had already been made in the Peninsula, but the late Emperor's great prototype was not properly reproduced until the second battle of Manassas. Jackson's command manoeuvred boldly around Pope's right flank and attracted his attacks on that side, while Lee brought up the remainder of the army and launched it frontally against what had now become the Union left flank. Thus Jackson's part of the battle was a defensive against greatly superior numbers, whereas Lee's was conceived as a crushing offensive against a tired and distracted enemy. This combination represents the full emergence of sophisticated technique on the Confederate side, made doubly noticeable by the dismal incompetence of Pope's unblooded army.

After Second Manassas there was a Union rout which almost bears comparison with what had happened at the same place in the previous year, and it may be attributed to the selfsame cause – an inexperienced Union army. On the other hand, this army *was* experienced enough to provide itself with an effective rearguard, and with fresh forces hastening to its support there was a successful evasion of the Confederate pursuit. This at least was an operation of war which had been mastered by Lincoln's lieutenants.

Having dealt with both McClellan and Pope by making attacks against the odds, it was natural that Lee should now be tempted to press his offensive still further. This time he ventured north of the Potomac itself, and gained a bridgehead in Maryland about the middle of September. On Antietam Creek, however, he found that his 'Napoleonic fan' of detachments had become overextended in the face of an efficient counter-concentration by McClellan. Lee was forced onto the defensive by odds of more than two to one, while his outlying sub-units hastened to the sound of the guns. He adopted a well conceived position in depth around Sharpsburg, although it had weaknesses due to the proximity of the River Potomac in the rear and the searching batteries of enemy rifled artillery on the heights in front. Nevertheless, he need not have worried. The Union attack was no better co-ordinated than it had been at First Manassas, and Lee was able to take a leaf out of the Duke of Wellington's book by exploiting the rolling terrain to conceal his men. The arrival of mobile reserves clinched the matter, and McClellan retired hurt. The Army of Northern Virginia had once again demonstrated its skill against a more numerous and better equipped opponent, although in retrospect Confederates did feel that they might have suffered fewer casualties if only they had bothered to improve some of their positions with the spade.

In desperation Lincoln now replaced McClellan by the bewhiskered Burnside, an inventor of patented small arms and one of the more forceful Union commanders at Antietam. Hampered neither by the indecision of McClellan nor the rash optimism of Pope, Burnside was tipped to make a better and more scientific job of the grand attack on General Lee than had either of his predecessors in command. By this time the Army of the Potomac had gained maturity and, despite the many setbacks, its ranks now included many who could fairly claim the status of veterans. Their armament had also improved, and they were now doubtless slightly heartened by their strategic success in forcing Lee back from Maryland to a position just west of Fredericksburg on the Rappahannock. It was here in December that the next battle was fought.

Unfortunately for the scientific Burnside, December was not really the best time of year to fight a battle in which the front-line troops (and their wounded) were expected to lie out under enemy fire overnight. Nor was it wise to choose the most open and overlooked spot in Virginia to assemble massed assault waves in front of a well entrenched and well prepared enemy. A range of less than one and a half miles might also perhaps have helped the supporting artillery to achieve its best effect, and with hindsight we can now see that it was a mistake to fight the battle with a wide river only 500 yards behind the front line. It was a pity, too, that about 50,000 assault troops were crammed into a frontage of less than a mile. This did not exactly help, but at least Burnside successfully extricated almost 90 per cent of his force at the end of the battle. Only some 13,000 casualties were lost from an army which deserved to be completely destroyed. Such is the true measure of Burnside's genius.

After Fredericksburg Burnside hit on the remarkably original idea of outflanking Lee's positions instead of attacking them head on, but in the event he became stuck in the mud on the bottomless roads and had to retire before he had even made contact with the enemy. He was speedily replaced by the colourful General Hooker, while the Confederates lay back in their winter quarters and continued to enjoy a fairly uneventful season. They were still as heavily outnumbered as ever, but it did not take much depth of perception for them to realise that they were confronted by an army which knew but little of the art of war. They may perhaps have reflected that a stricter application of Lee's offensive policy might have brought them still richer pickings at Fredericksburg, but on the other hand such a counter-offensive over open ground would have involved much greater risks[11] than an attack in the more familiar woods and swamplands of the Chickahominy. In this particular battle General Lee had surely not needed to take risks, because Burnside had already done so much of his work for him.

When Hooker took command of the Army of the Potomac he had the good sense to wait for the spring before he made his attack, and he was also shrewd enough to perceive the deep philosophical advantages of Burnside's second, 'outflanking', plan over the earlier 'bull in a china shop' version. When he came to put it into practice, however, Hooker found to his astonishment that Lee was not to be taken by surprise with such a predictable manoeuvre. The attacker was himself attacked and put at a considerable psychological disadvantage.

When Hooker crept stealthily through the woods around Chancellorsville he hoped to strike the rear of Lee's army while it was still in position overlooking Fredericksburg. He envisaged a formal, set-piece battle with reversed fronts, and was therefore unprepared to receive a heavy assault while he was still entangled in the Wilderness woodland. He hesitated and was lost. Lee and Jackson had admittedly taken some astonishing risks in attacking such a numerically superior enemy and in splitting their own force into no less than three separate fractions. Jackson's characteristic flank attack was a particularly bold stroke which ought, by rights, to have incurred the piecemeal destruction of the Rebel army. But, as in the Peninsula, the battle of Chancellorsville was not decided by rights and, although Jackson himself was accidentally killed at the very hour of victory, the mere fact of a Confederate attack was enough to demoralise the Union commander. His army was still in good fighting trim, but Hooker called off the battle and retreated.

As after Second Manassas, Lee now felt that the balance of morale between the two sides was such as to justify an extension of the offensive principle into the Union itself. His 'Napoleonic fan' began to move forward again and penetrated deep into the untouched foraging grounds of Pennsylvania. At Gettysburg on 1 July the Confederate vanguard bumped the enemy and triggered a closing of the fan around that point, although the absence of most of his cavalry did deprive Lee of much vital intelligence. During the following two days he was surprised to find that the Army of the Potomac – now under the unflamboyant but highly methodical General Meade – consistently kept just one step ahead of his moves. Each time the Confederates decided on an attack, they failed to set it in motion in time to forestall a Union reinforcement at that particular point. There was much slackness about the higher co-ordination of the Confederate attacks – the habits of the Peninsula once again – although to offset this failing we find that Meade was content to hold his defensive positions without attempting either a counter-attack or fancy manoeuvres – the habits of Beaver Dam Creek and Malvern Hill once again.

At Gettysburg Lee discovered that his formula for demoralising the enemy by making attacks against the odds had finally outrun its usefulness. Unlike McClellan, Pope and Hooker, Meade did not panic when his well equipped and well fortified army heard the Rebel yell rolling up and down the line. Instead, Meade gritted his teeth and tried to keep on functioning as carefully and systematically as the circumstances allowed. That he could do this at all sets him somewhat above the majority of his predecessors, but that he could do it in the particular circumstances of Gettysburg – when the Union faced real defeat for the first and only time in the war – speaks volumes for his tenacity. Conversely, the fact that he failed to seize his moment by launching a crushing counter-attack excludes him from the very highest echelon of generalship, especially as he allowed the Confederates to hold their positions until Lee was good and ready to move out.[12] The Army of the Potomac had not been transformed into a confident and nimble-footed counter-puncher by Gettysburg, nor was it ever to become one. But at least Gettysburg showed that there were limits to its capacity for self-inflicted damage.

What followed was a period of imbalance, or stasis, between the two sides in Virginia – a period of nine months without a battle, despite several attempts to achieve one, in which neither side felt confident to push home against the other. At first, in late July, while Lee was retiring to Culpeper, Meade made a clever but ultimately unsuccessful effort to cut him off at Front Royal. Then Lee tried to manoeuvre around Meade's right rear in October, but was outmarched to Centreville and did not press the attack. On the rebound he was finessed out of the Culpeper position and pushed back to the Rapidan, again by some clever Union manoeuvres; but when in late November Meade had finally brought his army face to face with the Rebels on Mine Run the determination to make an assault was still lacking. The strength of Lee's fortifications and the severity of the season persuaded first Warren and then Meade himself that it would be better not to go ahead. They retired to take winter quarters just as Lee was bracing himself to launch a counterblow. That blow never landed, and it was not until May 1864, with the arrival of Ulysses S. Grant as Union General in Chief, that operations were resumed in Virginia.

The fascinating period of 'non-battles' and shadow-boxing between Gettysburg and the Wilderness shows that both armies had now attained a high level of competence in the art of operational manoeuvre, but that neither felt sufficiently confident in itself to wish to precipitate combat. The generals had learned caution and the men had to some extent lost the ardour which they had shown at their peak, in the second year of war. Recruits were becoming harder to find, a high desertion rate had set in to stay, and the Union was experiencing a number of draft riots. Yet at the same time there was a certain resurgence of Northern confidence, since Gettysburg had proved that Lee could be held at least to a draw and the capture of Vicksburg showed that progress in the Western theatre was still being sustained. Although he might not view the prospect of battle with enthusiasm, the Northern fighting man now knew in his heart that he was on the winning side.

For the Confederates, by contrast, the ebbing tide of the war served to emphasise the need to conserve whatever scanty resources were available. Lee was finally persuaded to abandon his policy of speculative assaults, because it was now too costly in the face of the huge Northern war machine. Instead, he turned to a policy of digging in. He would confront his enemies with powerful fortifications which would either deter them from attacking or would give them a mauling if they did attack. In the event Meade usually thought better of it and backed down but when Grant arrived from the West he reverted to the idea of pushing forward. After a month of fearful losses, however, he too learned caution. For the final phase of the war there would be few convincing attempts to launch an offensive in the East, with the single spectacular exception of the last campaign of all.

Where Grant did differ importantly from his predecessors was in his idea that an assault, once launched, should be maintained for as long as humanly possible. He saw that McClellan and Hooker had accepted defeat prematurely, while their armies were still in good shape, simply because the Confederates had refused to accept that *they* were defeated. Grant therefore resolved that he would not himself fall into this trap, so when he followed Hooker's path into the Wilderness near Chancellorsville he did not stop fighting until there had been bloodbaths there, then at Spotsylvania, then on the North Anna, and finally – most costly of all – at Cold Harbor. This process of attrition was seen as especially spine-chilling and ruthless by contemporaries, and it did have the desired effect of driving Lee back to the fortifications of Richmond and Petersburg. Ironically, however, this led to a strengthening of those key

positions just at the moment when a detached Union force under Butler
had a good chance of storming them against only light opposition. Grant
might have done better not to have fought his battles after all! As it
was, he successfully manoeuvred his main army to the Petersburg front,
but failed to take the position by assault. His own men were now so
tired, and had suffered so many casualties, that the attack was soon
called off.

What followed was a ten-month siege of Petersburg which has impressed many commentators by its similarities with the trench warfare of
the First World War. A line 53 miles long was eventually established by
February 1865, with a garrison of little more than 1,000 Confederates per
mile (as opposed to about 15,000 per mile at the battle of Fredericksburg
only two years earlier). The wonder is that this line was able to hold out at
all, although it must be admitted that it was rarely tested seriously.
Perhaps the severest test had come at the battle of the Crater in July 1864,
but that attack – like so many others in this war – was badly mishandled
and the chance was lost.

The first three years in the East had seen the Americans learning some
very similar operational lessons to those learned by the French during the
first fifteen years of their Revolutionary and Napoleonic wars. Both had
started with untrained, not to say anarchic, armies: and both had gradually
evolved an effective system of march manoeuvres. The Americans
managed to do this faster than the French for a number of reasons – not
least of which must surely be their awareness of the French experience
itself – but the end result showed a remarkable family likeness. Even the
'non-battles' of late 1863 can be matched by numerous 'non-battles' in the
Spanish Peninsula and elsewhere. Yet by 1864 we find the emergence of a
different line of development, something which a Napoleonic soldier
would not have recognised so easily. This was Grant's protracted battle of
attrition from the Wilderness to Cold Harbor and its converse, the
institutionalised 'non-battle' at Petersburg which we today call 'trench
warfare'. We will return to this phenomenon at a later stage and attempt to
identify the tactical reasons for its appearance.

THE WESTERN THEATRE

While the Army of the Potomac and the Army of Northern Virginia were
locked in their epic struggle in the East there were other campaigns – of
ever-increasing intensity – taking place in the Western theatre. These had
a quite different character from the sustained manoeuvring over the
hundred-mile glacis between Washington and Richmond, especially because of the vast geographical sweep of the western territories themselves.
From Richmond to Vicksburg (by no means the furthest point of military

interest) was some 900 miles by rail. From Fort Donelson to New Orleans was almost 600 miles by steamer, and from Atlanta to the sea was over 200 miles on foot. It was little wonder that the Western soldiers had wider horizons and a more rugged approach than their colleagues back east. They were fighting a very different sort of war.

Much has been made of the differences between Western and Eastern theatres, and indeed there were many. It might be worth our while to consider, however, that in Napoleonic terms it was rather that the Eastern theatre of the Civil War had exceptionally small dimensions than that the Western theatre was exceptionally large. In 1805 the Emperor had marched some 750 miles from Boulogne to Austerlitz, and in 1812 he advanced about 600 miles from his start line to the high water mark of his invasion of Russia. It was therefore not the distances in themselves that made the Western theatre so exceptional or difficult to the Civil War campaigner, but the problems of resupply which those distances involved.

In most of Napoleonic Europe, apart from Russia and Spain, there was relatively dense settlement and therefore a relatively abundant supply of provender for man and beast. In the Virginia of the 1860s there was also a reasonable availability of supply – from the Shenandoah valley, from the Carolinas, from Maryland and Pennsylvania or, for the Union at least, by sea from still further afield. In the Western theatre, by contrast, the density of settlement was low, and therefore the foodstuffs available locally to an army were usually very limited. Only if a large volume of goods could be brought forward regularly from a base depot could an army stay in the field at all. Yet to achieve this there was little which could be done by wagon trains, since they were too slow and consumed too great a proportion of their own payload to keep going for long. For any large operation it was essential to stay near a rail or river line along which the power of steam could be harnessed to the army's logistic needs.[13] It was this one factor more than anything else that determined the course and direction of the Civil War in the West. It was this factor that dictated that the troops would be less well armed than their colleagues in Virginia, less well cared for, and generally less ready to forgo the temptations of looting.[14]

Much has also been written about the racial, or at least cultural, differences between eastern and western man – differences which are alleged to be quite as marked as those between the Northerners and Southerners themselves. Thus, whereas the 'typical' Eastern Federal is imagined as a Boston bank clerk – a tidy-minded urban teetotaller – and the 'typical' Eastern Rebel is imagined as a Virginian from one of the better families, the Westerners of both sides are seen as big-boned, loose-limbed farm hands.[15] There is a no-nonsense health and robustness about them which admits no role for the closed theoretical systems (whether

41

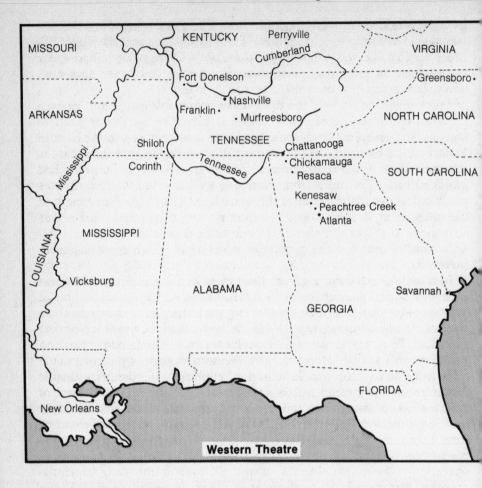

Western Theatre

actuarial or genealogical) of the Easterners. There is none of the deference to outdated European values, and none of the unnecessary ritual beloved by the hidebound professional soldier, either.

On this last point, in an unconscious echo of General von Moltke, one Union captain, Charles H. Slater, privately expressed the telling opinion that 'the Western rebels are nothing but an armed mob, and not anything near so hard to whip as Lee's well disciplined soldiers.'[16] The same opinion was repeated in another form by a Virginian who went west to join General Braxton Bragg's command:

The army is very badly generalled and the result is there is much demoralisation and want of confidence. Bragg ought to be relieved or disaster is sure to result. The men have no faith. The difference between this army & Lee's is very striking.[17]

Against this we have some similar evidence for the state of Sherman's

Union Army in November 1863:

This army looked quite unlike our own that had originally been a part of the army of the Potomac. They all wore large hats instead of caps; were carelessly dressed, both officers and men, and marched in a very irregular way, seemingly not caring to keep closed up and in regular order. These were faults in marching which we had been taught to avoid.[18]

Obviously drill and military punctilio were not among the strong points of Westerners, although an alternative explanation might perhaps be that it was the officers who lacked the necessary motivation to insist upon such things. This suggestion in turn leads us to ask just how military experiences in the West may have differed from those in the East, and how the balance of morale may have been formed by a different set of influences. It is this question that we will tackle now.

In 1861 Western operations were conducted on a very small scale compared with those in the East, and it was only in the following year that the main-force campaigning began to roll. When it did begin, the differences from the Eastern campaigns became dramatically obvious at once. This was not 'Napoleonic' manoeuvring at all, but steam-driven riverine warfare. It was also warfare in which the Union had the upper hand from the start. The pattern was that a relatively weak Confederate garrison would attempt to hold a fort on bluffs commanding one of the great rivers – the Mississippi, the Cumberland, the Tennessee – but would be unable to prevent a Federal flotilla from concentrating in the vicinity, landing a greatly superior force of soldiers, and attacking from the landward side. In these circumstances there was little the Rebels could do, since they were forced to spread their forces thinly over many far-flung outposts, while the North could keep its forces well concentrated and choose its moment to strike.

Inevitably, the vital first battle in the West was a resounding Union victory – so resounding, in fact, that the very expression 'unconditional surrender' was coined there by the Union commander, Grant. This was at Fort Donelson on the Cumberland in February 1862, and it was followed by a deep penetration up the Tennessee to Pittsburg Landing and Shiloh, near the important communications node at Corinth. Both sides at once moved all available forces to the spot – the Confederates hoping to surprise and overwhelm Grant by a major concentration, the Federals hoping to reinforce his exposed salient before it was too late.

In the event the Confederates did achieve their surprise, on the morning of 6 April. They successfully overran several Union campsites which, despite fears that they would be fortified, proved to be quite open to attack. Nevertheless, the Rebels failed to annihilate their opponent before his reinforcements arrived from the river. After 36 hours of confused woodland fighting they had been pushed back from the ground they had gained and were forced to retreat. Their commander, the promising Albert Sidney Johnston, had been killed and some 24,000 other casualties of both sides littered the battlefield. This had not been a big battle in terms of the numbers engaged, but it had certainly been a big one in terms of the percentage losses.

The Union forces now had a clear run to Corinth, although in compensation for their near-fatal failure to fortify at Shiloh they insisted on digging in carefully after every advance of two or three miles. It was General Wager 'Old Brains' Halleck who was responsible for this faintly ridiculous method of pursuing a defeated enemy, and it must be admitted that he could and should have made much faster progress. Nevertheless, we do have to remember that he had been the central organising genius behind the dramatic Northern successes in early 1862, and he ought to be forgiven his little idiosyncrasies. He was a classic example of a commander who is better at higher strategic direction than at leading armies in the field.

Fort Donelson, Shiloh, Corinth and a number of lesser riverine outposts that fell about the same time were all ultimately Halleck's victories. When taken together, they constituted a devastating blow to the Confederacy both in terms of controlling the strategic geography of the area and in terms of the morale ascendancy seized by the Union in this theatre at the start of campaigning. It did not matter very much that Halleck was now promoted out of his controlling position in this theatre, or that the Union offensive now suffered a series of checks and frustrations. The crucial fact was that the Confederacy had failed to achieve a psychological hold over its Western enemies, by early victories, in the way that it had done so brilliantly in the East.

In the autumn of 1862 the Confederates stabilised their position on the Mississippi by holding off the Union forces at Vicksburg and in the bayous north of New Orleans. They launched counter-attacks on the middle Tennessee and – most significant of all – started a policy of deep raiding which was to be a characteristic feature of the Western theatre for the remainder of the war. These raids were usually entirely cavalry operations, commanded by such hard-bitten railroad wreckers as Nathan B. Forrest and John H. Morgan. They exploited the overland mobility of the horse to compensate for their loss of riverine mobility to the US Navy. Nevertheless, on a number of occasions deep raids were also attempted with forces of all arms moving at the speed of the man on foot.

The first great example of an all-arms raid was Bragg's thrust into Kentucky at the head of some 35,000 men. In October he met a superior Union army around the water-holes at Perryville, defeated it narrowly, but then discovered that he had failed to secure a sound logistic base. He could not afford to hold his ground without a railroad to his rear, so he retreated back through Tennessee to the vital communications centre at

Chattanooga. From there he moved up to Murfreesboro on Stones River, to act as a shield against any Union advance from Nashville.

General Bragg was a great believer in offensives. He had seen Grant's offensive work well at Fort Donelson and his own almost work at Shiloh. At Perryville the tactical offensive had won the battle even though the strategic offensive failed. When he was threatened at Murfreesboro, therefore, he naturally moved into the counter-attack and did not stop until he had all but swept the enemy from the field. In a battle very similar to Shiloh, the Confederates overran several lines of Union troops among a tangle of cedar thickets. Also as at Shiloh, they lost their impetus and co-ordination when they reached the culminating point of their offensive. Renewed fighting two days later failed to restore fluidity to the battle, and Bragg withdrew in disgust. In this exceptionally bloody contest each side suffered casualties amounting to approximately one-third of its total strength.

Bragg's retreat from Murfreesboro cannot be compared to the Confederate retreats from Perryville or Shiloh, however, since on this occasion he had no real reason to abandon the battlefield to his foes. He succumbed to a failure of will, rather than of logistics or of numerical odds. It was comparable to McClellan's withdrawal in the Peninsula or Hooker's at Chancellorsville, and the reasons were probably similar. When two opposing armies are both hard pressed it is the side with inferior basic morale that will back down first.

In 1863 the focus of attention in the West shifted to the epic siege of Vicksburg and its subordinate operation at Port Hudson. Vicksburg finally fell on 4 July, after which the centre of action swung back once again to the line of the Tennessee. As always in the Western theatre, these far-flung operations required considerable advance planning and co-ordination, due to the times and distances involved. Logistics were of paramount importance, and the various cavalry raids against enemy supplies took on an unexpectedly great significance.

At Vicksburg itself there had been a long series of unrewarding frontal assaults and attempts to bypass the fortress by land and water. Eventually, in May, Grant did successfully insert a force to cut off the city from the east, with a series of dazzling offensive manoeuvres which kept the enemy in ignorance of his intentions in every phase. This operation showed him to be one of the most competent exponents of the mobile battle to emerge from the war, and it allowed him to starve out the Vicksburg garrison within two months. His attempts to accelerate the process by direct attack were nevertheless a failure. Whereas Grant had succeeded with tactical attacks against isolated detachments of the enemy army, he failed against the well fortified main body. Vicksburg eventually fell to starvation, not to frontal assault.

By September General William S. Rosecrans, the 'victor' of

Murfreesboro, had finally advanced through Chattanooga and was attempting to catch Bragg's army in an ambitious sweep of the 'Napoleonic fan'. Unfortunately he spread his fan too wide in thickly wooded terrain towards Chickamauga and – as at Murfreesboro itself – fell foul of a numbing Confederate counter-attack. The initial Union discomfiture was even greater than it had been at either Shiloh or Murfreesboro, and although there was a tenacious rearguard stand by General George H. Thomas, 'The Rock of Chickamauga', Rosecrans retreated hastily to his depot at Chattanooga. He was lucky that Bragg once again failed to press home his advantage, although this did not save Rosecrans his command. Grant and Thomas now took over responsibility for this sector of the Union line.

In the period of 'phoney war' which followed Chickamauga we find that the Confederates, although highly successful in the battle itself, were stalled by the Chattanooga defences and lost the initiative to their foes. They dug in on Missionary Ridge overlooking the town, and had leisure to contemplate the shaky state of their own supplies compared with the gradual build up on the Union side from late October. They also experienced acute problems of leadership, as Bragg attempted to make up for his failings of generalship by a harsh and unpopular regime of discipline. Within Grant's command, by contrast, there was a calm willingness to make good the losses in both men and morale which had resulted from Chickamauga. The underlying superiority of the Northern armies in the West allowed this rebuilding to proceed more smoothly than was generally possible in the East.

Grant launched his offensive in November, before the Confederates had managed to remove Bragg from the command for which he was by now so clearly unfitted. The Union attack possessed little subtlety and against the strong fortifications of Missionary Ridge it deserved to be roughly repulsed; yet by a near-miracle it actually became a devastating success. The further Grant's men ventured up the slope, the quicker the defences crumbled. The Rebel army was seized by a panic as great as anything seen so far in the war, and ran back helter-skelter to the old Chickamauga battle area. As an example of a central breakthrough of an enemy line by frontal assault, this battle could scarcely be bettered by a Blenheim, an Austerlitz or a Jena. Yet, as with so many other successful tactical offensives in the Civil War, there was a failure to follow through to an annihilating pursuit. Bragg's army was able to fall back and regroup because its rearguard performed well, because the Union lacked a sufficiently powerful and mobile *corps de chasse*, and because a small army in flight can move faster than a large army disentangling itself from a battle it has won. These were factors which reappear many times in the history of the war.

The battle at Chattanooga marked the start of a new era in the Western theatre, since the Union forces had now grown to such a size that they could counter the various Confederate raids, large and small, upon their lines of communication. They were even starting to mount a series of raids on their own account. With the elimination of the Rebel presence on the Mississippi and the defeat of Bragg, furthermore, the main armies of the two sides both recognised that the tide was flowing heavily in the Union's favour. Intensified attacks by the Union and intensified caution for the Confederates seemed to be the logical policies for the army to pursue in the spring campaigns of 1864. No longer did the Western Rebels have faith in the offensive principle, as they had so often in the past; it was scarcely an accident that J. E. Johnston, the cautious commander from Seven Pines, was now appointed to take command over them.

Johnston was a believer in the systematic and mobile use of fortification. If the enemy looked as though he would outflank one line of prepared positions, Johnston would manoeuvre to place a new line in his path. In the protracted movements from Chattanooga to Atlanta between May and July he applied this technique even more thoroughly than did Lee between the Wilderness and Petersburg at around the same time. Admittedly Johnston was helped by the rougher terrain and less numerous roads in his area of operations, but his main assistance came from the Union willingness to play a stand-off game. For a while the manoeuvres became a minuet between two armies which were both anxious to avoid a battle. For such sustained operations in such close proximity to the enemy there were remarkably few casualties.

Simultaneous with Bragg's replacement by Johnston there had been a major reorganisation of the Union high command, as Grant was called to the East and General William T. Sherman was promoted to command in the West. Sherman was very much a member of Grant's school and had distinguished himself in many of Grant's battles. Yet he was also a commander whose experience of the tactical offensive had rarely been a happy one. At Vicksburg at New Year 1863 his attack had been especially roughly handled, and at Missionary Ridge he had encountered far stiffer opposition on the left flank of the Union attack then had Thomas in the centre. When he started the operations to Atlanta he was therefore naturally chary of repeating such unfortunate experiments. Only very occasionally, as at Kenesaw Mountain on 27 June, did he attempt a frontal assault on the enemy entrenchments – and was duly repulsed. Even so the Kenesaw battle was relatively small in scale when compared with some of the frontal assaults in the East; there were only some 3,000 Union casualties as compared with about four times that number at Fredericksburg or six times as many at Spotsylvania. Obviously Sherman was a Johnston to Grant's Lee, when it came to the matter of ordering frontal assaults.

On the other hand it would be wrong to put these manoeuvres to Atlanta in the same category as the various indecisive 'non-battles' in the East during the second half of 1863 or during the Petersburg siege, since in the case of the confrontations between Sherman and Johnston there were some clear winners and losers. Sherman *did* make headway by his gradual and methodical technique, and Johnston *was* forced to give up ground. By 17 July the Confederates had surrendered some 80 miles of real estate and had fallen back across the Chattahoochee River to the outskirts of Atlanta itself. They could take some satisfaction, perhaps, that Johnston's army was still very much a force to be reckoned with in what was by now a highly unequal struggle, but many of them expressed a deal more dissatisfaction that he had given away so much without a fight. In these circumstances it was felt that a change was again required. A more aggressive policy than Johnston's was preferred, so he was replaced by General John B. Hood, the much-amputated doyen of the bayonet charge.

All Hood's early experience of Civil War fighting had consisted of successful assaults. In skirmishes in July 1861 and then during the retreat from Yorktown the next year, in the bloody but ultimately triumphant charge at Gaines's Mill, in the final attacks at Second Manassas and in some of the hardest combat at Antietam, Hood was always in the thick of it at the head of his unstoppable Texans. Nor did he learn caution, as so many others had done, during the middle war period. At Chickamauga he commanded the spearhead of Longstreet's decisive attack until he was wounded at the moment of success. There cannot have been a Civil War commander who made so many successful assaults, so it is little wonder that he had fretted and intrigued against the cautious policy of Johnston. When he took command of the army at Atlanta he was more than ready to launch it forward in a series of desperate offensives.

With the benefit of hindsight we can now see that these attacks were misconceived, leading to no result apart from heavy casualties which the Confederates could ill afford. Sherman's men were not to be shaken in the same way as some of Hood's earlier adversaries had been. They merely tightened their stranglehold on Atlanta and goaded their opponent into ever more desperate expedients. Hood's reaction was to go raiding against Sherman's line of communication. Although his own means of resupply were tenuous in the extreme, he chased north into Tennessee, aiming for Kentucky. At Franklin his assault was badly co-ordinated and failed as seriously as any of those before Atlanta, although the outnumbered Union forces did retire to Nashville. There in the middle of December a reinforced Union army launched a counter-attack which effectively destroyed Hood's army. The last major Confederate offensive had failed.

THE END OF THE WAR

While Hood was making his unsupported charge into Tennessee, Sherman was starting his own great raid into the deep South. Carrying everything he needed with him or foraging it along the line of march, he progressed through Georgia against negligible opposition. At Savannah he drew supplies from the fleet, then resumed his advance into the Carolinas. A trail of twisted rails and burning buildings marked his progress wherever he went. Whenever the improvised local militias tried to make a stand against his 60,000 seasoned veterans, they were quickly brushed aside with scarcely a battle.[19] Even in the biggest such clash, at Bentonville in March 1865, the combined US and Confederate casualties barely exceeded 4,000 men – a far cry from the massive butcher's bills of so many other Civil War combats. There was no disputing the fact that by this stage Sherman could come and go in Dixie as he pleased.

Meanwhile, in the East, Grant had prepared a spring offensive for 1865 which was to become the most brilliant of the war. Using a heavy spearhead of cavalry under the dashing General Philip H. Sheridan, he succeeded in turning Lee's right flank and forcing him to evacuate Petersburg and Richmond. In the scramble which followed Lee seemed to have lost his customary sureness of touch. His army became disjointed and hesitant. It found itself outmarched by elements of the Union forces which were large enough to impose delays and rerouteing. Finally on 9 April 1865 Lee was cut off by Sheridan near Appomattox Court House, some 75 miles to the west of Petersburg. Before the road block could be cleared the Union cavalry had been reinforced by heavy columns of infantry, leaving the Virginia Confederates in a cauldron from which there was no escape. They surrendered, and were followed nine days later at Greensboro, North Carolina, by the forces which had been facing Sherman. The war was effectively over.

Some authorities have identified no less than 10,000 different battles and skirmishes in the Civil War,[20] of which maybe fifty to a hundred can be considered significant in scale. That makes an undeniably impressive total for a four-year period, at least when measured against Napoleonic norms. We gain an impression that the Civil War was fought at a higher level of intensity than most earlier campaigns, with each soldier taking part in a greater number of combats. This must surely have offered exceptionally good opportunities for both leaders and led to learn their battlecraft by actual experience, and to build upon it by the observation of new situations. When we read some of the memoirs from the Virginia theatre we can find individuals who have undergone the ordeal by fire as many as a hundred times.[21] Such men can be accounted veterans indeed.

Yet the other side of the same coin is the sheer exhaustion which such intensive and continuous operations must have produced. For three years in Virginia the armies were vouchsafed an annual respite of only two or three months in winter quarters. For much of the remaining time they would be marching, fighting, or manning front-line positions. Excessively hot in summer and excessively cold in winter, the climate of the war zone would add its own particular discomforts and dangers to those contrived by the enemy. Only the strongest of men could survive this way of life for long, and those who did must of necessity have become expert in the art of minimising the rigours of their ordeal.

It has been found in many wars that fresh troops enter a campaign with plenty of enthusiasm but little skill. Then as they mature their skill improves until they reach a peak of efficiency, perhaps for their third or fourth battle. Beyond that, however, their skills continue but their enthusiasm and energy fade away. They get to know too much about the terrible risks which combat entails, and they take deliberate measures to keep out of it whenever they can. They stand appalled at the memory of their own lack of caution in earlier battles, while they were still green. They lose their sharp cutting edge and degenerate into the sort of veterans who do the minimum of what is asked of them, but no more.

The same evolution may be traced very easily in many of the Civil War regiments which were sent into battle so often in these crowded years of conflict. Whereas in 1862 and well into 1863 there was still usually enough collective enthusiasm to produce fluid battles of great ferocity, by the end of 1863 a significant part of this impulsion had been lost. The problem was not so much that demoralisation was rife – although there were indeed some notable examples of that, as in Bragg's army on Missionary Ridge – but rather that the veteran regiments had learned caution. Thus on Mine Run, and again at Cold Harbor, the Union troops were found to be writing their names and addresses on scraps of paper and pinning them to their uniforms, in order to permit notification to their next of kin in the event of their deaths. This forerunner of the modern 'dog tag' may doubtless be portrayed as a sign of high professionalism among soldiers who had 'seen the elephant' often enough to understand the practical problems which combat casualties involved; but on the other hand it may also be interpreted as a sign that these soldiers really did expect an imminent death.[22] In the specific tactical circumstances of Mine Run and Cold Harbor this expectation was certainly based upon a well informed assessment of the actual probabilities.

Then again we must remember the extraordinary caution of the skirmish battles from Chattanooga to Atlanta, and from Atlanta to Bentonville. With the minor exception of Kenesaw Mountain and some major exceptions during Hood's period of command, this campaign was fought on amazingly 'bloodless' principles. In the East the same thing was happening at the same time. The 'non-battles' between Gettysburg and the Wilderness, and then the ten months of inaction at Petersburg, showed that the armies had very definitely lost their taste for recklessness. Grant's month of frontal assaults from the Wilderness to Cold Harbor ought perhaps to be seen as an anomaly as unusual, in its way, as Hood's fortnight of rampaging around Atlanta. In both cases these offensively-minded commanders seem to have been acting against the ingrained ideas of their own forces, and they both soon gave it up to try some other policy. We gain an impression that it was the inertia of the veteran armies themselves that imposed a vital dampener on offensive action in both theatres, no matter who commanded them.

This is not to say that all the battles were always indecisive, for we have seen that they were not. Nor would it be true to suggest that all assaults invariably failed. As late as Spotsylvania or Nashville we find some excellent examples of successful frontal attacks. At no point during the war could it be said that the offensive idea had been completely abandoned, and the final campaign to Appomattox was itself a splendid affirmation of just that principle. Yet we can perhaps see that by the last eighteen months of the fighting many officers and men had lost their early uncritical acceptance of orders to make attacks. They needed firmer assurances than before that their sacrifices would produce a valuable result before they would venture into heavy fire. Too many successful tactical attacks had been left unexploited in past battles for the men to invest automatic trust in their commanders.

In all the above we can detect a learning process at work at several different levels. Inside the regiment there was an evolution from green-horn to hard-bitten veteran, or 'old lag', taking place more or less rapidly according to the specific circumstances of the particular regiment – its combats, its leaders, its marches and its losses. Then in each theatre there was an evolution of army morale according to the local experience of victory or defeat. In the East the Confederates enjoyed an impressive superiority over their opponents from the first, which began to be eroded only in 1864 by the underlying Union certainty of final victory. In the Western theatre, however, the Rebels made a much less happy start and were dogged by a series of unfortunate incidents, of both their own and their enemy's making. They never could shake off these disadvantages, and the Union effectively maintained the initiative through much of the war.

At the highest level of command there was a third learning process at work, as each army commander developed his own personal style, his own set of reflexes when faced by the great questions of attack or defence, advance or retreat. Some, such as Sherman and Johnston, learned from early experiences of failure in the attack that they should proceed more cautiously and use manoeuvre or entrenchments instead. Others, such as Hood and Grant, had sufficient early successes with the offensive to be convinced of its value for the remainder of the war. Lee had also learned this lesson at first, but found that he had to unlearn it after Gettysburg and follow instead a system similar to that of Sherman and Johnston. Yet other commanders again were more timid than any of these, and would creep forward with entrenchments when they ought to have been charging forward in headlong pursuit. Halleck at Corinth and McClellan at Yorktown are obvious examples, and Meade also sometimes betrayed related symptoms.

This diversity of experience and outlook among the Civil War commanders should warn us against leaping to hasty conclusions about the supposed obsolescence of the offensive. If some commanders refused to give it a fair trial while others enjoyed some good successes with it, we have room to believe that it remained an effective instrument when wielded by the right men. Certainly the European experience between 1859 and 1871 was highly encouraging for the offensive: in Italy, Denmark, Austria and France there were four quick and decisive campaigns which were all won by the attacker.

In the Civil War we should also remember that the balance of numbers and morale, or at least one of the two, usually weighed against the attacker. By their strategic circumstances the Confederates were often obliged to attack against superior numbers, while at least in Virginia the initial demoralisation of the Union troops was to cast a long shadow over many of their numerically strong offensives. There was thus a relative absence of overwhelmingly powerful or sustained attacks delivered against a weak opponent, of the type which might have produced results as brilliant as Napoleon's best. The Europeans of the Second Empire period, using essentially similar technology, were more fortunate than the Americans in this respect.

Finally, it may be worth noting that the Civil War included few clearly overwhelming attacks that produced such scanty results as did so many in the First World War. There would appear to be little good reason for making a link between the tactics of the two wars on this account.

2

Command and Control

A jingling staff was galloping hither and thither. Sometimes the general was surrounded by horsemen and at other times he was quite alone. He looked to be much harrassed. He had the appearance of a business man whose market is swinging up and down.

The Red Badge of Courage, p. 38

After the balance of morale and the preconceptions of commanders in the great questions of attack and defence, one of the most important determinants of a battle's outcome was the ease (or otherwise) with which a general's intentions could be translated into action. The techniques of command and control at his disposal might make all the difference between his effective intervention in the action or his being compelled to watch helplessly as it unfolded according to its own unaided momentum.

In Napoleonic times there had been some obvious limits to the extent to which a commander could direct the combat. The speed of communication was no higher than that of a horse, while the science of staff work itself was still very much in its infancy. Yet to set against this is the fact that the armies had been able to manoeuvre in close order around relatively small battlefields, well under the eyes of their generals. If an attack went wrong, as d'Erlon's did at the start of Waterloo, there was a good chance for a commander to try something different an hour or so later, or to rally the defeated troops and send them in a second time within, say, six hours. The Napoleonic assault therefore enjoyed considerable flexibility, as did the defence.[1] In both cases there was scope for cut, thrust and parry – even with formations as big as an army corps – within a very short time scale.

By the early twentieth century the exercise of command and control on the battlefield had undergone some important changes. Communications had been speeded by the telegraph, telephone and radio. Professional staff officers were at last available to see that all worked smoothly, or at least according to a system. Aerial observation had come of age and added a third dimension to warfare, while trains and trucks were revolutionising overland transport. Yet at the same time the difficulties were also multiplying as the armies grew far bigger and more cumbersome, with heavier *matériel* to move around with them and more complex problems of coordination whenever they wished to bring it to bear. Communications were liable to be cut within the range of enemy field artillery, and an attacker's manoeuverability across a bullet-swept and cratered landscape was reduced almost to vanishing-point – even though a defender retained his ability to move up reserves quickly from outside the battle area. An imbalance of mobility thus occurred, which deprived the generals of much of their ability to control events.

Where did the Civil War lie in this spectrum between Napoleon's relatively easy command and control and the more cumbersome conditions of the Western Front? In this chapter we will try to find out.

STAFF WORK AND THE GENERALS

At the governmental and higher strategic echelons of command the Civil War saw as much improvisation as in any of its other branches.[2] In the North the civilian Lincoln took a 'layman's crash course in strategy' and set about designing an administrative structure that would allow him to control and direct his generals, instead of forcing him to accept their views without question. By 1863 he had been largely successful in this, energetically abetted by the sinister Edwin M. Stanton – a lawyer, telegrapher and believer in 'business principles', who was Secretary of War from January 1862. Until the final promotion of Grant as General in Chief in the spring of 1864, however, the result of Union centralisation was effectively to diminish the authority of each local commander, from the Chief of Staff right down to each of the (excessively numerous) generals commanding departments. Admittedly, the North did have a bigger basic problem of co-ordination than the South, but nevertheless one suspects that the complex tensions of Washington politics were too often allowed to stand in the way of military rationalisation. It is hard to disagree with the suggestion of one recent commentator that 'Perhaps Lincoln was so completely a political creature that he simply did not know how to operate outside the context of party politics'.[3]

In Richmond, by contrast, there was an almost diametrically opposite mixture of administrative vices and virtues. Jefferson Davis was a military man and had been a successful Secretary of War in the 1850s. He succeeded rather better than Lincoln in co-ordinating the action of his generals and in harmonising the military with the political requirements. Yet his achievement was essentially personal, and he failed to establish the type of modern administrative machinery which a Stanton would have demanded. Confederate bureaucracy remained creaky, old-fashioned and riddled with restrictive practices which the Southern war effort could ill afford. Although the armies and their commanders could be placed in the field rather more effectively than those of the North, their logistic supplies, communications and supporting staffs never were.

When we turn from the higher direction of the war to our more immediate concern – battlefield command and staff work – we find a picture which varies widely in scale and scope according to the particular level being considered. Whereas a seasoned infantry brigade might contain no more than 1,000 troops deployed on a frontage of less than 500 yards, an army could easily exceed 70,000 men spread over many miles of front. The brigadier could thus hope to see most of his troops at a glance, and command them almost instantly by riding back and forth from one regiment's colonel to the next; yet for the army commander it was unlikely that more than a small sector of the battlefield could be visually observed, or that personal leadership could be exercised over more than one or two important subordinates in any period of two or three hours. Commensurate with his high responsibilities, furthermore, an army commander would have a sprawling headquarters organisation to supervise, including experts in many specialised branches such as administration, filing, military law, signals and logistics, not to mention engineering and gunnery. The army headquarters became a major unit in its own right, requiring appropriate management skills to work properly. At Gettysburg Meade's army headquarters numbered no less than 3,486 souls, exclusive of the engineers, the artillery reserve and the headquarters of the individual army corps.[4]

Headquarters establishments were certainly much less lavish at the lower echelons of command, and they could be deployed correspondingly nearer to the front line. At the level of the humble brigade the provision described by Sherman as ideal amounted to just an adjutant, a quartermaster, a commissary, a couple of medics and a pair of well-mounted young ADCs. When it came to controlling a battle only the adjutant and the ADCs would have a role – the former as the brigadier's right-hand man and 'general expediter', the latter as his link with higher commanders or outlying detachments.[5] This amounted to a minimal staff apparatus, and it is clear that a brigadier was not expected to play a very complicated role. If he could give personal supervision and tactical advice to his colonels, then he would have done his job. The more ethereal functions of staff work could be left to his superior division and corps headquarters.

A major problem, however, was that the techniques of staff work remained rather too personal – essentially verbal – even at those higher levels above the brigade where they should have been more formalised. Following Napoleonic practice, the giving of orders was intended to be verbal in the first instance and then confirmed in writing, but in the Civil War the lack of sufficient trained staff officers too often meant that the confirmation was left unwritten.

One example of this occurred at the Battle of Corinth on 3 October 1862, when a serious Confederate attack was unleashed against Rosecrans's Army of the Mississippi. Once he had identified the enemy's main thrust

Rosecrans sent his acting chief of staff to give new orders to General Hamilton, the commander of one of his four divisions, who had been posted to guard a flank but who had not thus far been engaged. This was quite a complicated verbal order which called for a speedy movement by Hamilton's force around the enemy's flank and rear. Rosecrans believed that it might well have secured a spectacular victory, if only it had been carried out as promptly as intended. Unfortunately, however, Hamilton either misunderstood or mistrusted the order, and refused to budge until he received it more clearly. Since the officer sent to relay it could not offer any useful elucidation and had not written it down, he had to go all the way back to Rosecrans for the written version which should have been provided from the start. As a result the vital moment was lost and the Confederates escaped with only a moderate defeat, as opposed to a decisive beating. Rosecrans summed it all up as follows:

Hamilton's excuse that he could not understand the order shows that even in the rush of battle it may be necessary to put orders in writing, or to have subordinate commanders who instinctively know or are anxious to seek the key of the battle and hasten to its roar.[6]

To which we can merely add that a more succinct and pointed summary of the Prussian general staff ideal would be extremely difficult to formulate.

The improvised Civil War armies suffered badly from their lack of trained staff officers. This was particularly true in the early part of the war, and probably more true of the Confederate armies than of the Union. There was a marked tendency for generals at all levels to appoint their assistants on a more or less arbitrary or *ad hoc* basis, often with family or mildly political considerations weighing as heavily as military ones. Nor was there any reliable source from which a steady supply of qualified men could have been drawn, since neither side possessed a staff college or staff corps.[7] The only school was the bitter one of hard experience, and it was only by a painful process of trial and error that the higher functions of command eventually came to be exercised properly in the Civil War armies.

For the Confederates not even Lee could find a satisfactory answer to this problem until at least the summer of 1864. In the Peninsular campaign of 1862 his staff work had left a very great deal to be desired and, although there had already been some improvements by Second Manassas, his attacks were still suffering unnecessary delays and failures of co-ordination even as late as Gettysburg.

The staff-work problem usually had a more serious effect upon offensive than defensive tactics. It added a sombre background shading to all too many of the Confederate attacks – in the West as well as the East. Shiloh, Chickamauga and Atlanta were all battles in which faulty liaison helped deprive a promising offensive of decisive results.

On the Union side there was initially an equal failure of staff work. Grant allowed himself to be surprised at Shiloh, just as did Hooker at Chancellorsville. McClellan at Antietam badly failed to co-ordinate his attacks, while Burnside at Fredericksburg was positively elephantine in his lack of tactical finesse. Only by the time of the Wilderness campaign in 1864 can we identify a really well organised and self-assured staff which, despite its rather haphazard and unplanned origins, could bear comparison with any in the world. Ironically, however, there were many other factors at work in the summer of 1864 – not least Lee's perfection of staff work in defence – which robbed the Northern attacks of victory. Grant found himself in the unfortunate position of one who has at long last mastered a particularly recalcitrant part of his problem, only to find that the easier parts have now suddenly become more difficult.

Both sides in the Eastern theatre suffered a rather disappointing end to the war, as far as staff work was concerned. In the Appomattox campaign the unexpected restoration of mobility, after ten months of siege, took everyone by surprise. Orders went astray, conflicting chains of command interfered with each other, and golden opportunities were wasted. Lee's loss of operational mastery is clear enough from the result of this campaign – but Grant's perhaps is less so. Nevertheless, General Chamberlain, who marched with the spearhead V Corps and finally commanded the surrender parade, has left us a bitter catalogue of complaints against the way the Union staff work was managed. Three competing authorities – Grant, Sheridan and Meade – each had somewhat unclear responsibilities. On the night of the White Oak Road battle this meant that V Corps HQ received no less than five sets of orders, none of which was compatible with any other.[8] During the battle of Five Forks Sheridan rushed around making *ad hoc* command appointments and dismissals as the mood struck him, often without knowledge of the true situation, and often without informing those most directly concerned. Chamberlain and others from the solid old Potomac Army were shocked at this cavalier and non-regulation manner of proceeding, although they were also spellbound by Sheridan's personal charisma. Like a Murat or a Rommel he could lead from the front; but like them also he was no kind of a staff officer.[9]

Having said all this, it is worth remembering that not even Wellington and Napoleon had commanded perfect staffs, and some serious failures of co-ordination may be detected in even their most brilliant operations. Both of them also lacked regular organisations for staff training and indoctrination, just as they in turn failed to leave lasting staff institutions behind them at the end of their wars. In many respects we may therefore say that the Civil War commanders were little worse off than their Napoleonic predecessors with the possible exception of the Prussians – although even the great Prussian general staff was to achieve its finest triumphs only after 1865.

The personal nature of leadership in the Civil War can hardly be stressed too much. Not only was relatively little written down, but much of a commander's understanding of what was happening depended on direct personal experiences, especially the direction and intensity of the noise of firing.[10] A commander would want to know whether he was hearing a smattering of carbine fire from a cavalry screen or the regular volleying of a formed line of infantry; whether it included a few light cannons or a full heavy battery; whether it was building to the climax of a charge or settling into a desultory long-term exchange. Such 'combat impressions' as these were identified by the Germans in the Second World War as still being important to a headquarters, even then; but they had been far more important in the 1860s.

Bugle calls provided another vital aural clue to the ebb and flow of combat, and each brigade had its own identifying signal.[11] Just before Chancellorsville Hooker had also provided the Army of the Potomac with distinctive corps and division cap patches, to stimulate unit pride as well as assist identification on the battlefield. Even more important, perhaps, were the flags. Each regiment, each division HQ and each corps commander had a distinctive flag which could be read off by a skilled observer. Admittedly there was considerable latitude for error in identification – but at least the basic clues were systematically provided for all to see.[12]

On the other side of the coin battlefield liaison was badly hindered by the heavily wooded terrain in which so many actions took place, since trees reduced visibility, smothered sound and generally made it difficult for generals to 'read the battle'. They also made it difficult to navigate when – as in Lee's army in the Peninsula in 1862, or Grant's in the Wilderness as late as 1864 – there were no good maps available. Even at the best of times maps would be in short supply, and a considerable effort always had to be put into sketching and distributing new ones. In the trackless regions in which the armies frequently manoeuvred it was really quite surprising that anyone succeeded in finding their way at all.[13]

The problem of maps was particularly acute for couriers, whose task would often be to deliver messages at top speed to a named individual who was moving around unpredictably within an unknown and uncharted area. The only possible technique to locate him would be to ask constant questions of those encountered *en route*[14] – a process which could yield some highly misleading or even dangerous replies. Many were the errant ADCs who blundered accidentally into an enemy force in the course of searching for friends. Many too were those who lost their lives in the attempt to deliver a letter – for example, at Corinth two orderlies were killed in quick succession in the course of Rosecrans's blighted correspondence with the tardy Hamilton.[15]

Generals could also blunder into the enemy unexpectedly, either with merely themselves and their immediate staffs or – more seriously – with their entire commands.[16] At night, in mist or smoke, or simply when uniforms and flags were not prominently displayed, there were many opportunities for mistakes, especially if commanders were leading from the front or had neglected the essential arts of scouting. The prudent commander would therefore take considerable precautions when he ventured near the front – for example, Sherman tells us how he would creep up carefully to occupy viewpoints in the safety of the 'little forts' of the skirmish line.[17] Nevertheless, the time and deliberation involved in such measures must surely have reduced the general's ability to react quickly and flexibly to a changing situation. There was an apparently insoluble dilemma between command effectiveness in a mobile battle and the personal safety of the commander.

These difficulties often struck Civil War soldiers as exceptional and novel. One of Meade's staff officers complained that: 'the typical "great white plain", with long lines advancing and manoeuvering, led on by generals in cocked hats and by bands of music, exist not for us'.[18] He had perhaps heard that Napoleon's battles were all conducted in this manner, and that at least one of them had even been fought on a real parade ground.[19] The contrast is unfair, however, since quite a high proportion of Napoleon's battles had actually taken place in difficult terrain such as forests, mountains and towns. In any case, his armies had generally been larger than those of the Civil War, creating a greater intrinsic problem of control, and they had far fewer maps to go round. Napoleonic generals were surely no less likely than those of the Civil War to lose themselves or their couriers – or to mistake the nationality of friends and enemies alike. They experienced the same dilemmas of whether leadership should be from the front or from the rear, and they suffered from some equally inexperienced staff officers. In none of these respects were the two eras all that very different after all, although it must at least be admitted that the Civil War generals no longer used a cocked hat to support their still-gaudy ostrich plumes.

In common with his Napoleonic predecessor, of course, the Civil War commander was also subject to his quota of human failings. This was especially true in the early years of the conflict, before quality had risen to the top and serious incompetence had been hounded out or assimilated. Nevertheless, right up to the end there were still plenty of petty jealousies and frictions between individuals, as between Halleck and Grant after Shiloh or Johnston and Hood before Atlanta. There was no shortage of vain posturings, bombastic rantings or spiteful intrigues. There were fits of uncontrollable temper and bouts of suicidal gloom. Sherman's manic glorying in destruction was never quite entirely sane, nor was Jackson's bleak Calvinist self-righteousness and secrecy. Above all, the influence of the whiskey bottle upon military operations can be traced throughout the generalship of the Civil War, recorded in lovingly censorious detail as for perhaps no other war in history.[20] We sometimes feel that the earnest New Englanders who sponsored this conflict regarded strong drink as an enemy second only to the institution of slavery itself.

The foremost general of the Republic admittedly suffered his moments of drunkenness, and from time to time fell off his horse. Nevertheless, he did win the war and was elected president twice over at the end of it. Equally, the personality clashes, intrigues and unjust policies of so many other Civil War officers doomed them to fall far short of the absolute standards of efficiency which might have been attained in a perfect world. Yet all this by no means proves that overall performance was any worse than the international par. We should remember that in these matters the marshals of Napoleon had set the standard at a shockingly low level – and even they had managed to win more than their fair share of combats.

BATTLE-HANDLING

When we look at Civil War concepts for controlling a battle, we find that they, too, are strikingly Napoleonic in character. The commander-in-chief selects an approach road for each division or corps, and at a given point along this route either he or they decide that the moment is ripe to deploy into line of battle. With luck their flanks will rest on a natural obstacle or another friendly formation. If this is not the case there may have to be some hasty reshuffling until a satisfactory position is achieved. The important thing, however, is to get a continuous line of men into place facing the enemy. In column of route or in scattered detachments the troops are desperately vulnerable to attack, but once the line has been formed they will be able to deliver their full fighting power. The deployment of a sound battle line is thus the central principle upon which all tactics are ultimately based.[21]

In this first phase of action the commanders at every level will seek to align their men onto neighbouring units and onto the terrain. Fields of fire will be chosen, batteries sited and plans formed for the action which lies ahead. A short harangue with wise words of advice may be given to the soldiers, if there is time, or at least a few crisp orders to their officers. Then the attack will sweep forward towards the enemy, still trying to maintain its linear formation, or if on the defensive the line will stand its ground until the situation changes for good or ill.

A great principle of both attack and defence in the Civil War was to flank the enemy with fire or movement, or both. A defensive position would be stronger if it was laid out to give cross-fire onto the enemy's avenue of approach, and it would in turn become vulnerable if it could be enfiladed by hostile artillery.[22] Considerable ingenuity would therefore often be expended, if time permitted, in establishing the most favourable layout in this respect.

In the attack there was an equal realisation that flanking movements were preferable to frontal assaults. This had been the doctrine of Frederick the Great a century earlier, and it was enshrined in every textbook.[23] Indeed, it was almost a geometrical inevitability in linear warfare that a flank march would be favoured by any attacker who wished to concentrate the bulk of his force against a fraction of the enemy's. The idea would be to leave only a small part of one's own force facing the enemy frontally, well protected by fortifications. He could beat vainly against this bulwark, distracting himself from the main body creeping around his flank, until the trap was ready to close.

The theory of the flank attack lay behind the Union plan at First Manassas, at the start of the war, and it was finally used to prise Lee out of his Petersburg position at the very end. In between, there had been every possible variant on the flanking attack, from Burnside's ill-fated Mud March and Meade's abortive Mine Run operation, to Longstreet's spectacular charge at the end of Second Manassas and his less timely attempts to turn the Union left at Gettysburg. In the Wilderness campaign Grant won fame for his 'jug handles', while in the West Sherman earned himself the nickname of 'flanker'. Even the fair sex apparently understood this much of tactics, if we are to believe the saucy story retold by one of Sherman's men:

one of the ladies of Georgia notoriety told one of our generals about our style of fighting, 'You'uns don't fight we'uns fair; you always come round on our eends'.[24]

There was a practical difficulty in arranging a flank attack, of course, because it had to move past the enemy's line before he could change front either to cover himself or to break contact altogether. Since he was likely to have chosen the terrain, and since he would be geometrically on the 'inside track', the advantages would all be his. Unless the attacker's army was considerably superior in numbers, furthermore, it might run a severe risk of counter-attack while winding its cumbersome way across the defender's front. It was for these reasons that many Civil War flank attacks failed to achieve a breakthrough, and why a direct frontal assault was sometimes adopted after all.

For the frontal assault the theory was to achieve surprise if possible, but even if it wasn't possible, to rupture the enemy by concentrating a succession of attacks in line, one after the other, against a relatively narrow sector of his position. Each attack would weaken the enemy, even if it failed to break him. If the process was repeated often enough he would surely crack in the end. This was very much a Napoleonic concept, lifted directly from French thinking, and it did sometimes work well.[25] Nevertheless, as we shall see, it was basically just as unreliable as the flanking attack itself.

The flank attack and the central assault in successive waves did at least both encapsulate a recurrent theme in the writings of Jomini and his followers, namely the principle of concentration. This doctrine had – very much so – been made in Europe, despite its subsequent Civil War appropriation by Nathan B. Forrest when he 'got there fustest with the mostest'.[26] In his semi-literate backwoods way, the founder of the Ku Klux Klan was merely restating an idea which must already have been in the minds of his better educated colleagues when they made their dispositions for combat.

So far we have examined the phase of the battle in which the general should still enjoy reasonably good control over his troops, provided his staff work has been efficient and the enemy has refrained from springing any surprises. The deployments will have taken place according to the drills for which the men have trained and the emotional atmosphere, although perhaps tense and highly charged, will not yet have prevented the units from being manoeuvred with almost the precision of chess pieces. When battle is fairly joined, however, all that will change in a flash. Units will move forwards or backwards without orders; they will ignore their alignments to disperse as individuals or to huddle together as a mass. Each part of the line will see the problem in a totally different light from any other, and attempt to apply its own solutions irrespective of the needs of the higher command. Tension will give way to terror, battle rage or numbness.

The second phase of battle-handling now begins. The general will seek to maintain his control of events by sending orders to correct the discrepancies which have developed, and he may even intervene personally to rally a shaky unit. Depending on his temperament he may shout himself hoarse with oaths or patriotic slogans, or he may sit back at his folding camp table and send a provost guard to make some arrests as an example to malingerers. He may jump lustily into the heart of the fight, or he may keep above it, cold and aloof. Whichever technique he applies, however, there will come a moment when he realises that the troops engaged with the enemy in the front line are effectively beyond his control. Their ordeal will have become so much their own affair that they can no longer be ordered forward into increased danger, nor be moved sideways or rearwards without the risk of a panic rout.

The most effective way to influence a battle which has reached this stage lies in the judicious use of reserves. The commander must turn to units he has so far held back, and commit them where they are needed most – to plug a gap that has opened in the line, to take an enemy in flank, or simply to reinforce a unit which has run down its stocks of ammunition and enthusiasm. The commander must make an assessment of just where and when his reserves can be used to best effect, and then he must ensure that they actually go there and do their job on time. It makes an apparently simple task and is easily stated in theory, but, as Clausewitz was at pains to remind us, 'in war even the easiest things are difficult'.[27] As we have already seen, this had been the experience of Rosecrans at Corinth when he tried to bring up Hamilton for a counter-attack on the enemy rear – and that was by no means an exceptional case.

The problem at the lower levels of generalship was not that the principles of using a reserve were misunderstood, for they were not. If we look at one of the first battles of the war, Seven Pines, we can see that a sophisticated technique was already being attempted by both sides. The Union camps were laid out in four fortified lines, with Casey's division providing the outpost line and the first battle line, about half a mile apart; Couch's division the reserve line a further half mile behind that; and two brigades of Kearney's division a mile and a half beyond, to act as a final longstop in case of extreme need.[28] Within each of these lines the 'reserve' principle was applied – for example, the outpost line consisted of skirmishers hiding in a wood in front of a line of abattis, supported by a picket reserve and an entire regiment behind that. During the battle this line was further reinforced from the rear by four regiments and a battery, before all were forced to retire in confusion to the second line. Once there, they were again reinforced from the rear and became intermingled in one formless mass until pushed back again onto the next reserves coming forward. So the battle went on.

On the Confederate side at Seven Pines we find an equally thorough understanding of the principle of reserves. General D. H. Hill started his attack on a cannon-shot signal with his division in two wings, each wing preceded by a skirmish screen of one regiment with a brigade in columns a few hundred yards behind. A quarter of a mile behind that came a support brigade or, in other words, a 'reserve' for the front-line fighting.[29] Unfortunately the terrain turned out to be too wooded and swampy for precise drill alignments to be maintained, and the front-line brigades were soon mixed up with the skirmishers. They became excessively bunched as they all tried to deploy on too narrow a front. At this moment the support brigade arrived in the middle of the battle and attempted to pass through to get at the enemy. In theory such a leapfrog was the correct way to relieve the front line, but in practice it led to still greater confusion:

the passage of lines being a feat in tactics which had never been practised by any of us, large fragments of those regiments who were left without field or company officers were joined in and continued forward with that brigade.[30]

The battle seemed to sag for a while, until Hill was able to persuade one of his brigadiers – not without demur – to take his command around a flank.

This had a disproportionate effect on the enemy and momentum was regained. At this point the division was reinforced further by a brigade detached from another division, and so the battle continued.

On both sides Seven Pines shows how reserves could be used to keep the fight going, in a sort of see-saw of fresh impetus coming from outside the seat of the battle. The tactical commander was forced to accept that his initial schemes and deployments soon fell to pieces under the enemy's fire, and his task was then a matter of finding some new units to begin it all over again.

Seven Pines also shows us, however, that even the best foresight and planning could still lead to fairly complete chaos. The major weakness in the use of reserves in the Civil War lay in the manoeuvre known to the drill sergeant as the 'passage of lines'. If this could be done crisply, a fresh and formed body could be presented to the enemy where only a few seconds before there had been an exhausted and disorganised rabble. This manoeuvre could often be made to work smoothly when relatively small units relieved a long-range or slow-moving firefight.[31] When it was a matter of large masses passing through other large masses under heavy close-range fire, however, it became a sure recipe for disaster. The disorder of the existing line would almost instantly contaminate the fresh troops and neutralise much of their potential value.[32] In fact there was only one thing that could disorder a formed body more quickly than a mismanaged passage of lines – and that was a successful enemy charge.

In Napoleonic times the problem of the passage of lines had been no less acute than it was to be in the Civil War, although with some of the better-quality units in both wars its dire effects were doubtless considerably reduced. The important thing was to attempt this manoeuvre only when the conditions were favourable, and never otherwise. Veteran regiments understood this well, and on occasion flatly refused to be reinforced when a fresh unit was sent up from the rear to help them.[33] But it is less clear from the records that commanding generals were particularly sensitive to this problem. Time and time again we find that their battle plans were based upon a succession of reserves arriving to a fixed timetable, irrespective of local conditions and innocent of specialised training for this highly complicated manoeuvre.

At least the commitment of reserves in Civil War battles was free of many of the difficulties which were to attend their use in the First World War. They did not have heavy baggage trains which needed days of planning to move, nor did they have a physical problem of marching across the battlefield to reach the fighting. They could be manoeuvred with relative ease in the attack no less than in the defence. Their only problem, unlike their latter-day successors, lay in passing through the lines of the troops already engaged.

So far we have discussed the use of a reserve at levels lower than the corps, and we have seen that it was a skill which was well understood in theory if not always practically applicable. When we look above the corps to the use of reserves at army level, however, we find a rather different picture. The army headquarters in the Civil War seems to have been just that little bit too far away from the local realities of combat for the correct management of reserves to have been easy or normal. At this level the state of the art in the 1860s seems to have slipped back from that of Napoleonic times to, perhaps, the flounderings of the 1790s. However well the Civil War division and brigade commanders may have understood the principle of reserves, it would appear that their army commanders, alas, did not.

At Antietam McClellan held back a whole third of his army and never used it at all, while at Gaines's Mill he fed in a second line too late to stem the flood of the Confederate advance across what should have been an impregnable position. At Fredericksburg, on the other hand, Burnside made the opposite mistake of squandering his massed reserves in an attempt to reinforce failure at the foot of Marye's heights; and at Chickamauga Bragg did something similar when he insisted on sustaining Polk's unsuccessful attack, when he might have reinforced Longstreet's breakthrough instead. In other battles too few reserves were held back at the start, with the result that commanders would hover anxiously around the rear edge of the battle area waiting to seize any tired and dusty column of men who happened to be passing, to send them into the thick of the fighting before they had time to draw breath. In some cases this predicament was the inevitable result of the Confederate overstretch in the face of almost double their own numbers, and at other times it was caused by an unavoidably imperfect view of the true situation. Nevertheless, on several important occasions it was a symptom of improvident forward planning.

Unlike Napoleon, the Civil War commanders do not seem to have been anxious to adopt the concept of a specialised reserve – a *masse de rupture* or *corps de chasse*. There was no equivalent of the Imperial Guard, designed to intervene at the crisis of combat as a powerful all-arms battlegroup. In Napoleon's thinking such a force had been intended to deliver a crushing blow at the psychological moment against an already weakened segment of the enemy's line. First the artillery would clear the way, then the infantry would seize the main position, and finally the cavalry would exploit beyond it. The 'intelligent co-operation of the three arms' would thus be an important part of an attacking general's duty – not merely in his preliminary dispositions for battle but also at telling moments selected during the climax of the struggle itself.

In the Civil War the 'intelligent co-operation of the three arms' seems to have remained something of a dead letter, at least until Sheridan's lightning campaign to Appomattox. As one artillery officer put it, 'The fact is that we have no general who has shown himself able to handle infantry, artillery and cavalry so as to make them co-operate together'.[34] There are several reasons for this neglect, such as the complexities of the terrain and the unfamiliarity of the concepts, but of particular importance was the organisational structure of the armies. At the start of the war each state had put forward its own regiments, legions and independent companies according to its own ideas about how such things should be arranged. Only gradually did the central governments succeed in subordinating this administrative anarchy to a more standardised principle in which the concept of a guard reserve might have had a chance to flourish. But even then there remained an unfortunate un-American ring to the 'Imperial' title which Napoleon had seen fit to append to his famous shock troops.

The process of concentration is most clearly seen in the artillery, where individual batteries were gradually grouped into battalions under one commander, and eventually into a single organisation for each army, under a chief of artillery empowered to act independently of the infantry corps commanders. Under the pressures of their predicament it was the Confederate Army of Northern Virginia which achieved this rationalisation most quickly, although even there it was only after a long campaign of trial, error and bureaucratic in-fighting. In the Union Army of the Potomac the road was longer and more beset with obstacles for General Hunt, the victor of Malvern Hill and champion of the Napoleonic conception of massed fire. It was only at Gettysburg that he finally won his point. In the Western armies, however, the ideal of massed artillery was rarely achieved by either side, and generally lacked formal institutional expression.

As far as the infantry was concerned there was never any serious attempt to mass the most experienced men in a single reserve formation, and although certain regiments and brigades did become particularly renowned for their high fighting qualities they were left dispersed around the army in a haphazard manner. Certainly there was little attempt to give special training as storm troops to selected units, although on at least one occasion, with Upton's brigade at Rapahannock Station and then Spotsylvania, this had produced impressive results.[35] Unfortunately, an even more deliberate attempt to do it a few weeks later turned into a complete débâcle. This was with Ferrero's 4th Division of Burnside's Corps at the Petersburg Mine Crater on 30 July 1864, when all looked set fair for a decisive breakthrough. At the last moment, however, the division was stood down on the grounds that it was coloured, and an unprepared white division was put in its place. This force proceeded to bungle the assault, the 4th Division was called in too late and in the wrong formation, and the whole corps suffered an extremely bloody repulse. The Battle of the Crater is acknowledged to have been potentially one of the most successful infantry attacks of the war, but it failed largely because the principle of a specialist assault division was overridden.[36]

When we turn to the cavalry we find a sorrier story even than with the artillery and infantry. All too little attempt was made to create a heavy cavalry reserve for use on the battlefield in the Napoleonic manner, and all too much attention was paid to those who claimed, variously, that firepower had made battle cavalry obsolete; that the terrain was unsuitable; that heavy cavalry was 'un-American'; that it was too costly, too difficult, too dangerous, or (if all else failed) logistically unsound.[37] Admittedly, there was a grain of truth in each one of these grumbles, but then much the same critiques could equally have applied to Napoleon's own heavy cavalry reserve. The fact was that where Napoleon had gone ahead and fielded a powerful force of this kind the Civil War armies baulked at the prospect. They did eventually assemble some impressive cavalry formations, but they failed to place them in the same relationship to the battlefield commander as Napoleon had done. Thus at Gettysburg Lee's cavalry was found to be too far away from the scene of operations to exert much influence, while on the Union side it never occurred to anyone that a massed charge of cuirassiers should have been the logical next move after the repulse of Pickett's attack, and might have converted Lee's local discomfiture into nothing less than the total destruction of his army.

With too little sure-footed staff work, an inadequate system for the passage of lines and no concept of offensive reserves at army level, it is scarcely surprising that Civil War battles were often determined by local expedients and improvisations rather than by the master-plans of commanders-in-chief. Although the lower echelons of the hierarchy probably possessed battle-handling skills equivalent to those of their Napoleonic predecessors, the fact remains that the higher commanders often did not. Hence the defending army would tend to enjoy a certain advantage over the attacker, since in conditions of low overall responsiveness and control it was the static occupant of terrain who could keep going longer than the active and creative challenger. Running a successful defensive battle imposed simply less of a strain on an army headquarters than maintaining a successful offensive.

INTELLIGENCE AND SIGNALS

There are two factors of great importance for battlefield command and control which we have not so far discussed – intelligence and signals. Even the worst-handled offensive could win victory if it could identify and strike an undefended enemy nerve-centre, while even the best-planned defence would crumble if the enemy's successes were reported too late for counter-measures to be taken.

At the strategic level the Civil War commanders were generally well supplied with intelligence, since there could be little true security in a war between two geographically intermingled populations sharing essentially the same language and culture. The Washington and Richmond newspapers were regularly traded over the picket lines along with such items as coffee and sugar from the North and tobacco from the South.[38] Prisoners of war were often garrulous and sometimes frank with their captors, all of whom could understand what was being said. Spies and scouts frequently slipped into enemy camps and cities, and experienced little difficulty recruiting informants resident behind enemy lines.[39]

When McClellan was led astray in 1862 by Pinkerton's grossly inflated reports of Lee's strength, it was paradoxically not a result of any lack of information. Pinkerton actually had very good details of the Confederate units in the field, but he made the mistake of imagining that he had not got the full picture. The great detective therefore invented additional units of his own, and fleshed out small regiments into big ones, with the result that the leaders of the Army of the Potomac came to believe they were outnumbered at least two to one.[40] This was certainly a major intelligence failure, but it was a failure of analysis and assessment rather than a failure to collect the raw material.

When it came to tactical as opposed to strategic intelligence, the basic facts became more difficult to collect. In sparsely inhabited woodland an enemy army could sometimes disappear for several days on end unless it was rather carefully watched. Even though its general intentions and order of battle might be well known to the 'friendly' commander, he could still be tactically surprised if he lost contact for even a few hours. Thus Pope had not expected Jackson to appear on his flank and rear at Second Manassas, nor did Johnston anticipate some of Sherman's jinking in the forests between Chattanooga and Atlanta. If an army or corps commander sincerely wanted to lose himself, apparently, he had a sporting chance of success.

Many techniques were used to keep the enemy army under surveillance, although none of them was ultimately foolproof.[41] Balloons were used extensively by the Union in the early part of the war, and could claim some successes; but they were eventually judged to be inadequately rewarding for the effort they required. Wire tapping was intermittently used, as were personal reconnaissances by generals, staff officers and a veritable menagerie of colourful characters proffering more or less dubious credentials as spies. As in so many other wars, the simple process of terrain analysis in the light of 'inherent military probability' could also unmask an opponent's schemes with surprising frequency.[42]

The major part of tactical surveillance, however, rested with three agencies. The first was the ordinary picket line routinely thrown out by an army as a screen and 'tripwire' against surprise. The pickets in the Civil

War were generally intelligent and sociable fellows and they often conversed with their opposite numbers from the enemy side, or at least kept their eyes open for changes in the tactical situation. Ruses such as 'Quaker guns', false campfires, and so forth, could sometimes deceive them, but not often for long. They were certainly the best authorities available for their own little patches of ground. The trouble was that they did not always see the need to report their findings to anyone else, particularly not in a systematic way. Even if they told their own superior officers, there was no guarantee that the important items would be relayed upwards to the agencies specialising in intelligence. For example, at Chancellorsville the Union XI Corps was locally aware of Jackson's manoeuvres against it, but the message failed to percolate to the army staff until the corps was already fleeing past in panic.[43]

The second major source of tactical intelligence was much more specialised. It consisted of the signallers posted at intervals along an army's front to man the semaphore stations or 'wig wags'. By virtue of their particular duties these men found themselves sitting on prominent viewpoints – armed with brains, powerful telescopes and a means of instant communication to army headquarters – for hour after boring, uneventful hour. If they scrutinised the enemy lines as well as their own neighbouring signal stations, that was not in the least remarkable. Nor was it surprising if they often reported useful snippets of information about what the enemy was up to. They were ideally placed to do professionally what the picket lines could do in only a haphazard and random manner.

Finally, it was the cavalry that bore the brunt of the intelligence effort in the zone of tactical interest beyond the army's direct line of sight. As in all previous and many subsequent wars, the cavalry possessed an ideal combination of mobility, responsiveness and fighting power for this role. Indeed, at some moments in the Civil War it almost seemed as though this was about the only role the cavalry *could* fulfil. It was certainly a role well suited to the somewhat irregular, individualistic, rough-riding assumptions of so many of the American horsemen. They could probe the enemy in small groups, set ambushes, attack minor detachments or flee from main bodies. They could indulge in the sport of mounted duelling when they encountered their own kind, or in the more profitable business of plundering when they found a succulent supply dump or convoy. With the increased availability of field-glasses, moreover, they could see further into an enemy camp than many of their Napoleonic predecessors had been able to.[44]

Inquisitive enemy signal stations could usually be identified and avoided by a force which wished to conceal its movements, and some complex deceptions were occasionally employed.[45] Alternatively, the signallers were so few in number, and so exposed to view, that they made a choice target for snipers.[46] Against cavalry the problem was less simple and, unless it enjoyed a significant numerical superiority in this arm, an army on the move ran a good chance of being identified and watched. At the start of the war in the East the Confederates had usually enjoyed the upper hand in mounted reconnaissance, but when Hooker took command of the Army of the Potomac he gave its cavalry scouts a much-needed boost. From then until the end of the war their effectiveness progressively increased. In the West the picture was less clear – and the more wooded terrain obscured it still further – but there too we can detect a gradual shift in favour of the Union as the war went on.

When we turn from intelligence to communications, we find some still more dramatic improvements over Napoleonic capabilities. The electric wire telegraph was of course the most significant and pregnant for the future, although its usefulness in the Civil War was largely strategic rather than tactical. The Union Secretary of War, Stanton, made sure that his ever-growing web of civilian long-distance lines took precedence over every other form of signalling, especially since he personally controlled the product at its centre in Washington. In military terms this was something of a disaster since it not only furthered the concentration of power in the hands of politicians but it was also a considerable setback for the army's own infant tactical telegraph. As was the case with certain other new weapons appearing in the Civil War, political and institutional pressures conspired to delay a revolution in battlefield technology.

The tactical telegraph was first sponsored and deployed by the indefatigable ex-surgeon Myer, the father of US military signalling.[47] As early as the summer of 1861 he had procured a 'flying telegraph' train and, although the first model proved unsatisfactory, he had a better design purpose-built by Beardslee of New York in time to see action during McClellan's Peninsula campaign the following spring. On this occasion the wooded terrain made the visual wig wag impractical, just as in several later battles fog or mist was to neutralise it. Hence the electric wire was seen to be the only realistic alternative to mounted couriers, and with eighteen messages successfully transmitted in two hours its battlefield début was acknowledged to have been convincing. In the static battle at Fredericksburg the Beardslee telegraph also worked well, although in the more fluid Chancellorsville campaign it failed badly. By this time the institutional pressures against the US military signallers were growing and, although some thirty Beardslee trains were eventually constructed, few saw field use. Myer was dismissed; his telegraph was taken over by civilians using less advanced equipment; and, despite the slow expansion of the tactical system until the end of the war, it remained Stanton's long-distance network that held the priority.

The tactical telegraph never did completely displace the mounted courier and the wig wag, even in the technologically advanced Army of the Potomac. By the summer of 1864 Grant was stringing wires down as far as division HQs,[48] but not even he could dispense with written messages for the local minutiae of battle-handling, or for the longer and more formal orders at higher level. In any case the telegraph had some artificial help from the exceptionally stately pace of the battles at this phase of the war, particularly the siege of Petersburg. There was time for a complex net to be established, as there had not been in many of the earlier combats. Nor did any other Civil War army quite scale the same peaks as Grant with the tactical telegraph: Lee suffered from a shortage of wire by the end of the war, and the Confederates in the West were rarely able to use this medium on the battlefield at all.[49] Sherman admittedly followed Grant in his enthusiasm for short local lines within his army, but it was still the strategic rear link which held priority for him. He, like all the other Civil War commanders, made full use of the more manual methods of signalling to supplement the electrical ones.

Perhaps inevitably, it was the less technically minded Confederates who took the lead in the use of couriers and wig wags. They had learned their importance early in the war, especially following their chastening failures of staff work in the Peninsula campaign. They formed a regular corps of signallers earlier than the Union, and maintained its integrity longer. Admittedly, both their tactical military and strategic civilian telegraphs remained far less developed than the Union's, and they also failed to break Myer's more advanced wig wag ciphers. Nevertheless, for much of the war their skill at passing messages across a battlefield was as good as, or better than, their opponents'.

It would be a mistake to underestimate the efficacy of the wig wag, which was actually a considerable step forward over Napoleonic technique. With flags by day and beacons by night it could be interrupted only by clouds, mist or screens of trees. Although Sherman was to complain that these had a knack of intervening precisely when important messages had to be passed, he did on occasion himself achieve some remarkable feats with the wig wag.[50] Nor was he alone. At First Manassas E. P. Alexander had used flags to warn of the Union assault, while at Gettysburg the mere existence of Captain Hall's signal station on Little Round Top imposed a significant delay on Confederate preparations to attack. At Antietam and Fredericksburg the Northern gunners even used wig wags for fire control at long range – surely a notable step forward in the state of the art.[51]

Higher commanders in the Civil War did generally enjoy more rapid means of battlefield communication than their Napoleonic predecessors – and in some cases even than their First World War successors. In this respect they could be said to have enjoyed particularly good command flexibility and responsiveness, although it was unfortunately counterbalanced by far less impressive attainments in some other areas of staff work and battle-handling. The problem tended to be not that commanders lacked information about what was going on, or that they were unable to transmit timely orders to their troops, but rather that they were often unsure of what policy to follow and unfamiliar with the general 'feel' of commanding an army in battle. In the French Revolutionary wars it had taken the best part of a decade for these skills to be learned, and even with their post-Napoleonic wig wags and telegraphs the Civil War commanders did well to make as much progress as they did within the four short years of their conflict. Nevertheless, the fact remains that the final balance of command capabilities remained only rather tentatively 'Napoleonic' in character.

Looking at the Civil War from the opposite end of the nineteenth century, we can equally state that the conditions of command and control had few points of contact with those of the First World War. The staff work was of an altogether lower order of sophistication, the battlefields were tiny by comparison, and the duration of combat far less. Effective leadership from the front was still a practical possibility, and communications still allowed a commander to manoeuvre large formations flexibly at short notice in both attack and defence. The battles were marked by a rapid see-saw as local reserves were committed in quick succession from opposite sides – a far cry from the cumbersome set pieces of the Somme or Verdun. Had the Civil War commanders been just a little more experienced they would doubtless also have mastered the use of large-scale all-arms reserves, thereby reaping more decisive results from their offensives than they actually achieved.

3

The Rifle

The flames bit him, and the hot smoke broiled his skin. His rifle barrel grew so hot that ordinarily he could not have borne it upon his palms; but he kept on stuffing cartridges into it, and pounding them with his clanking, bending ramrod. If he aimed at some charging form through the smoke, he pulled his trigger with a fierce grunt, as if he were dealing a blow of the fist with all his strength.

The Red Badge of Courage, p.81

Perhaps unfairly, our stereotype of the Civil War battles as 'indecisive' places less emphasis on command and control technique than upon the new rifles, particularly the Springfields and Enfields which were carried by almost every infantryman by 1865.[1] In this chapter we will examine these rifles, and others in use at the same time, and try to reach some conclusions about their distribution and effectiveness in battle.

THE VARIETY OF WEAPONRY

In Napoleonic times the infantry had been armed with a smoothbore muzzle-loading flintlock musket. At extreme range this was capable of firing a heavy round ball up to four or five hundred yards, although not even a remote chance of accuracy was conceivable at such a distance. A regiment firing massed volleys might hope for reasonable results at one or even two hundred yards in deliberate target shooting, but in battle it was found that the average soldier could realistically hope to hit the enemy only at fifty or sixty yards. In part this finding was based upon the poor target training, crowded ranks and the general air of excitement which the soldier took with him into combat, but in part it was also based upon the technical specifications of the gun itself. The smoothbore musket was an inaccurate weapon which was effective only at close range.[2]

Between 1815 and 1861 a number of important improvements were incorporated in the Napoleonic musket which went some way towards making it a more recognisably modern weapon. The flintlock was replaced by a percussion cap system of ignition, which gave improved wet-weather performance and a marginally higher rate of fire. More was learned about ballistics and the theory of aiming, with at least lip-service being paid, at

73

long last, to the idea that the soldier should be allowed to set his sights according to the range.[3] Most important of all, however, was the introduction of rifling combined with a cylindro-conoidal 'Minié' bullet. The resulting rifled, muzzle-loading, percussion musket, or rifle musket, could fire over a thousand yards at extreme range, with reasonable accuracy up to about five hundred. What is much less clear is whether or not the average soldier in combat actually obtained very much benefit from these improvements, since many of the same factors which had limited range and accuracy in Napoleonic times continued to apply throughout the Civil War. Fields of fire were often very short, the soldiers were generally unskilled in the use of their weapons, and the officers were anxious not to engage in indecisive long-range fire.

Despite the technical possibilities of good accuracy at long ranges, tactical theory still rested upon the idea of massed fire at close range. This has been condemned as obsolete or unimaginative by many historians, although such writers tend to forget that even with the greatly improved ·303 magazine rifle of 1914 the whole emphasis of musketry training continued to rest upon volume at close range rather than accuracy at long. The famous 'mad minute' of fifteen shots per minute, which at Mons was mistaken for machine-gun fire by the Germans, should therefore be seen as a linear descendant of Napoleonic practice rather than as a change to some new policy of training the rank and file as snipers.[4]

Snipers did exist in the Civil War, of course, just as they had in Napoleonic times. They were very thoroughly trained in the arts of long-range marksmanship, and were equipped with the finest rifles money could buy. In the 1860s this meant a perfected version of the rifle musket – perhaps an English Whitworth or Kerr, or maybe a heavier rampart rifle. Such weapons were often clumsy and slow to load, with delicate sights and specialised ammunition. They were inappropriate to massed use in the firing line, and were in any case available only in minute numbers. Nevertheless they did offer a practical possibility of effective aimed fire at ranges of a mile or even more, and thereby greatly extended the zone in which the infantryman felt himself at risk. It did not take very many hits, or even near misses, for this lesson to be learned. The dread of snipers was quick to spread throughout the armies of the Civil War to an extent which was quite disproportionate to their actual numbers and effectiveness.[5]

As we have already stated, however, the main line of development lay less with the super-accurate sniper's rifle than with the common soldier's weapon for massed volleying in a zone approximately a hundred yards deep. Even the humble smoothbore could give excellent results in this type of firing, especially when loaded with buckshot or 'buck'n'ball'. Such a load gave a spray of projectiles up to about fifty yards which was actually a lot more lethal than a Minié ball fired at the same range from a rifle. It had served the Americans well in their earlier wars – far more so than the vaunted Kentucky sniper's rifle – and by the 1830s it had also been adopted by European theorists.[6] We should therefore beware of assuming that the Civil War rifle musket was a greatly superior combat weapon to the Civil War smoothbore, even though their technical specifications were markedly different.

The really important break from Napoleonic weaponry dated from 1811, when the Hall rifle was put forward as a single-shot breechloading variant of the normal musket. This rifle fired a round ball, but could combine this with rifling because the breechloading system removed the need for ramming from the muzzle. The Hall could also claim treble the rate of fire of its rivals, which made it a 'massed volley' weapon *par excellence*. It was tested in 1813 and again in the 1820s, but despite some impressive successes it was never distributed very widely.[7] A major expansion in breechloaders had to await the improved technologies that were appearing in the 1850s. These, however, came too late for the mass production of such weapons before the Civil War had started. Only a few infantry regiments took breechloaders into action – mostly Sharps and Burnsides, but also a handful of the older Halls.

In the case of repeating rifles the technology came later still, while the war was actually in progress. These used a magazine of metallic cartridges – seven for the Spencer and sixteen for the Henry. After each shot the flick of a lever would bring the next round into the breech, allowing a rate of fire phenomenally higher than anything previously seen. Whereas the musket and the rifle musket might fire one to five rounds per minute, depending on the skill of the operator and the state of the weapon, the Spencer could be fired seven times in ten seconds and the Henry in one test managed 120 rounds in five minutes.[8] Far more than the single-shot breechloader, it was the repeater that represented a true revolution in firepower. Despite its many teething troubles and the intrinsic unreliability of its ammunition, it could pump out so many shots in such a short time that it offered a new perspective in tactical theory from that used by the old carefully aimed one-shot weapons. The repeater looked forward to the tommy gun rather than backward to the Napoleonic flintlock. To put it a little differently, we could say that whereas the improved rifle musket of the sniper had added a new spatial dimension to the battlefield of the Civil War the repeater had added a new temporal dimension to the close-range volley.

The wide variety of Civil War small arms makes a fascinating object of speculation for the military amateur, but for the quartermasters of the 1860s it must have been something rather worse than a nightmare. Not the least of their troubles was the infuriating habit of regiments to be unspecific when they submitted their indents. Thus it was little use to know that 500th Infantry Volunteers required 'ten thousand rounds of ammunition' if there were fifteen different types of rifle, musket or carbine officially listed for Confederate use – and in practice there were really far more than that – or twenty-four muzzle-loaders and twenty-seven breech-loaders listed for the Union. The Union listed more than twenty-five different types of cartridge and, once again, actually used many more than it officially listed.[9] To complicate matters still further, there were many regiments on both sides which carried more than one type of weapon – for example, at Gettysburg 1st Minnesota had a mixture of ·69 smoothbores, ·69 and ·58 muzzle-loading rifles as well as the Sharps breechloader. Here we see almost two centuries of small arms development concentrated into a single regiment. Admittedly, it was an exceptional case, although not quite so exceptional as one might imagine. In the same battle there were about ninety Union regiments (about 36 per cent of the total present) with more than one type of weapon. Most of these carried two types of muzzle-loading rifle of ·58 or ·577 calibre, so their ammunition was compatible; but in 38 cases (16 per cent of all the regiments) there were at least two different ammunitions.[10]

THE DISTRIBUTION OF WEAPONRY

Table 3.1 shows the total distribution of weapons in the Union regiments at Gettysburg. Although the majority of Union infantrymen did have an Enfield, a Springfield or something better, there was still, in July 1863, a significant shortfall from total modernity (even if we accept the Enfield or Springfield as an adequate standard of modernity, which some would dispute). In earlier battles in the war,[11] or in the Western theatre,[12] or in the Confederate armies, there was likely to be an even higher proportion of inadequate guns, since in general terms we can say that the Union forces in the East were the most richly furnished of all in the matter of *matériel*.

TABLE 3.1   UNION INFANTRY ARMAMENT AT GETTYSBURG [13]

Total Union infantry regiments	242	
Armed partly with smoothbores	16	6·5% approx.
Armed wholly with smoothbores	10	4·0%
Armed partly with 2nd-rate rifles	13	5·3%
Armed wholly with 2nd-rate rifles	26	10·5%
Total armed with substandard weapons	65	26·3%
Armed in whole or part with breechloaders	7	2·9%

(In all, 70·8% of the regiments were armed with Enfield ·577 or Springfield ·58 muzzle-loading rifles.)

76

According to General Alexander, the Confederates had started the war with only 10 per cent of their men armed with modern rifle muskets, while many of the recruits could be given no weapons at all.[14] Measures had to be taken to insist that volunteers for service brought their own arms, regardless of whether that meant a hunting rifle, a shotgun or a horse pistol, and utterly regardless of any sort of standardisation.

When Vicksburg fell in 1863, on the other hand, Grant tells us that he captured 60,000 good imported rifles – almost certainly Enfields.[15] This shows that Southern purchasing agents in Europe had done their work well, as had Southern blockade runners in the Caribbean. Their efforts were all the more impressive since Grant's own forces had until that time been particularly badly equipped themselves. At Vicksburg it had actually been the Confederates who had achieved a superiority of *matériel* over their opponents.

Lee's Army of Northern Virginia was apparently able to equip all its men with good modern rifles after the Gettysburg campaign, but the Confederates in the West were not so lucky. Although they had certainly risen above shotguns and the other early expedients by 1863, they still remained heavily dependent upon smoothbores and second-rate rifles. Table 3.2 shows a breakdown of their weapons in the Chattanooga and Atlanta campaigns. At least by June 1864 the Army of Tennessee was managing to move out of smoothbores and had almost completely jettisoned the notoriously inefficient Belgian rifle. None the less, the fact remains that these soldiers, fighting one of the most important campaigns of the war, were far from ideally armed.

TABLE 3.2   ARMAMENT IN THE ARMY OF TENNESSEE, AUGUST 1863–JUNE 1864[16]

Calibre	13 August 1863		19 June 1864	
	No. held	%	No. held	%
·577 Enfield/ ·58 Springfield	14,100	45%	27,108	56%
·54 Lorenz, Mississippi, etc., rifles	2,000	7%	15,841	30%
·69 M1817, etc., mostly smoothbores	12,000	36%	5,369	12%
·52/·53 M1804 rifle, Hall rifle, etc.	3,000	9%	779	2%
·70 Belgian rifle	900	3%	64	0·1%
Sniper rifles	?	?	75	0·1%
Spencer repeaters	?	?	58	0·1%
Other (shotguns, etc.)	?	?	10	0·01%
Total in Sample	32,000		49,303	

Beyond these indications it is unfortunately very difficult to find accessible evidence for the distribution of weapons in the Civil War armies. We are usually forced back on rather ambiguous snippets of information from individual regiments. Thus we learn that 14th Indiana in 1861 was issued 'only muskets' (presumably smoothbores) apart from its flank companies and five sharpshooters per centre company (who presumably received muzzle-loading rifles). On the battlefield of Antietam, over a year later, the battered remnant of this regiment was at last able to pick up enough rifles for everyone, to their great relief, and to jettison their obsolete muskets.[17] Another example is 13th Massachusetts, which started the war with a heavy and inefficient 'Winsor' rifle but was lucky enough to receive the Enfield a few weeks later, before it had left Boston. On arrival on the Potomac near Harper's Ferry it replaced a regiment armed with smoothbores, and at once astonished the Rebels in this sector by hitting them at ranges previously considered safe.[18]

Not all the Enfields distributed were perfect in quality, however, for when another Boston regiment, 35th Massachusetts, received its rifles at about the same time it found that

They were somewhat defective, the cones being too much case-hardened and quite brittle, so that a large number were turned into the Washington Arsenal within three weeks; nevertheless, so great was the scarcity of weapons at the time, the regiment was thought very fortunate to get them.[19]

When this regiment went into action shortly before the battle of Antietam it found that the problems with their rifles had not been solved, since

... the cones of some snapped off at the base, rendering such arms useless, for we had no tools to extract the stumps; and the ammunition seemed to fit loosely, so that some were disgusted when their bullets dropped into the water a few rods from the muzzle. Several men were found who had never fired a gun.[20]

The longer the war went on, the more likely it was for each soldier to have obtained a serviceable rifle. Thus in Hood's division there had still been several score unarmed men in each Texas regiment at Fredericksburg, but after the battle they were able to find abandoned guns in plenty.[21] At Fredericksburg the Confederates salvaged a total of 9,091 weapons, including 250 new Springfields, 3,148 'improved' M1842 muskets, 1,136 'altered' older muskets, 772 Austrian rifles, 78 Belgian and 42 Springfield muskets, 478 M1841 Mississippi rifles, 59 Enfields and no less than 13 flintlock muskets. The balance was made up of every conceivable type of small arm, among which the more modern and useful weapons were in noticeably short supply.[22] It would therefore seem that the Union forces had already swapped their own substandard weapons for newer types where they could, before they allowed the Confederates to have second pick – or maybe it was the Confederate scavengers themselves who handed in their old weapons while keeping the new ones they had found.

Battlefield salvage was an activity which the soldiers of both sides always pursued with ardour, especially when it came to money, footwear and weapons. 'Johnny Green of the Orphan brigade' tells us that at Shiloh his regiment 'swaped our very indifferent guns for their splendid Endfield rifles',[23] while J. H. Worsham of 21st Virginia reports that before Gaines's Mill his regiment had one company armed with Springfields, one with Enfields, one with Mississippi rifles and seven with percussion smooth-bores. As a result of the battle, however, the whole regiment got 'top class' guns. Nor did the salvaging process stop there, since in 1864 this writer became 'the best equipped man in the army' when he captured a 16-shot Henry repeater.[24] In the matter of personal armament, apparently, there was no end to the successive improvements which an enterprising soldier could make.

By the end of the war the Union was so well equipped that it could sometimes afford to use captured rifles to corduroy its roads, and even from the South we sometimes find reports of guns being so plentiful that they were allowed to go rusty in open-air dumps.[25] These stories, however, are probably less a reflection of any true strategic abundance in weaponry than an indication of the short-term administrative difficulties that the sudden and massive influx of rifles after a victory could create. The hard fact remains that for much of the war neither side could get its hands on as many good rifles as it wanted, and expedients like battlefield salvage were often an urgent necessity. This certainly applied to the Confederacy more than to the Union, and we may almost be tempted to attribute some of the South's famous love of the offensive to this factor. The side which bivouacked on the field after combat was, quite simply, the side which became better armed for the battles which followed.

Salvage was not, of course, the only source of weaponry for the Civil War armies. Apart from the highly inadequate pre-war stocks, both sides could draw upon purchases in Europe as well as their own domestic production during hostilities. The Union was especially successful in converting its home industries to the mass production of armaments, and it more than made up in quantity for what it lacked in the quality of the deals it made with European governments – who were naturally very willing to modernise their arsenals at American expense. Because of their shaky industrial base and financial status the Confederates lagged far behind in both these fields, but even they somehow managed to find enough arms to keep going, after a fashion.

This is a complex subject in which there are all too many difficulties in collecting, comparing and assessing evidence. But, as a general guess at the shape of procurement, Table 3.3 may perhaps be some help to the student in a hurry.

TABLE 3.3   SMALL ARMS PROCUREMENT BY CENTRAL GOVERNMENTS [26]

	Held pre-war	Bought in Europe in war	Home production in the war	Total
CONFEDERATE				
Smoothbore	140,000	40,000	3,000	183,000
Rifled, muzzle-loading	35,000	300,000	100,000(?)	435,000
Breechloading				
single-shot	—	—(?)	4,000	4,000
Repeaters	—	—	—	—
Total shoulder arms	175,000	340,000	107,000	622,000
Revolvers	—	3,500	13,500	17,000
UNION				
Smoothbore	400,000	100,000	10,000	510,000
Rifled, muzzle-loading	100,000	1,000,000	1,750,000	2,850,000
Breechloading				
single-shot	3,000	—	300,000	303,000
Repeaters	—	—	100,000	100,000
Total shoulder arms	503,000	1,100,000	2,160,000	3,763,000
Revolvers	6,000	14,000	400,000	420,000

Note:

These figures are broad guesses and must not be taken to imply precise numbers. They are intended only to show an order of magnitude. They also exclude both battlefield salvage and procurement by individual states (as opposed to procurement by Washington DC or Richmond Va.) It has been estimated that the Confederates salvaged perhaps a quarter of a million small arms on the battlefield, and that the individual Confederate states imported a further quarter-million from Europe. For the Union the figures must have been at least comparable.

Nor is a distinction drawn between 'first' and 'second' quality rifles, or between small arms for infantry and those for other types of troops. Thus the astonishing US production of smoothbores in the war was almost entirely composed of cavalry carbines, whereas the smoothbores obtained from other sources were mostly infantry muskets. Equally, the rifle muskets bought in Europe by both sides ranged wildly from the incomparable ·451 Whitworth target rifle to the execrable ·70 Belgian rifle.

Among other things, it is clear from the statistics that the Confederacy was never a serious contender in the field of breechloaders, let alone repeaters. Although a good number of these weapons were captured at various times, the specialised ammunition for them was all but unobtainable. It was only the Union that possessed the industries which could provide the wherewithall. Yet the Union manufactured less than half a million of these guns, perhaps 12 per cent of their total procurement of shoulder arms, and of these the vast majority went to the cavalry. The Union infantry was thus almost as fully deprived of the new weapons as were the Confederates.

There are many reasons for this imbalance which need not concern us here. Suffice it to say that the accepted philosophy of the US Ordnance Board before 1865 was that an infantry weapon should be reliable, robust and 'soldier proof'; it should have long range, good accuracy, a powerful projectile and (in order to reduce ammunition consumption) a relatively low rate of fire. When the Enfield or Springfield muzzle-loader was compared with any breechloader available during the war, it fared better in every single one of these tests. It also had the advantage that it was familiar to and trusted by its users, and that the government armouries were fully conversant with its manufacture.[27] Finally, it performed far better in terms of its cost, since the breechloader was a great deal more expensive to manufacture. This indicated not only that breechloaders would impose a greater strain upon the taxpayer – in wartime such factors were not seen as supremely important – but, more important, that they would require more time and labour to produce. If we think of the dollar cost of these weapons (given in Table 3.4) as an index of their difficulty of manufacture, we may begin to see why the Ordnance Board preferred to produce a large number of the older weapons rather than a small number of the new. Any other policy might have left a lot of Union soldiers without any arms at all.

THE SOLDIER AND HIS AMMUNITION

The normal armament for each man was a rifle musket, a bayonet and a full pouch of ammunition – that is, four packs of ten rounds each. For much of

TABLE 3.4 APPROXIMATE COSTS OF CIVIL WAR WEAPONS [28]

(Normal prices paid at the time in US currency.
Note that a soldier's monthly pay was $11–13, plus rations.)

A smoothbore muzzle-loading musket	$1–10
A rifled muzzle-loading musket	$10–20
A single shot breechloader	$20–30
A repeating breechloader	$37–65
A sniper's rifle or rampart rifle	$60 +
A revolver	$12–25 (but up to $65 in certain cases)
A machine gun	$1–2,000

the war the Union included a number of Williams self-cleaning cartridges in each pack of ten rounds, although these bullets did have somewhat inferior ballistic properties to the Minié ball (or 'Minnie', as it was popularly known).[29]

For 'special occasions', when a battle appeared to be likely, the soldier would often be issued additional ammunition. Thus for the Wilderness battle the Union troops were given ten extra rounds, just as they were for the battle of Culp's Plantation on 22 June 1864.[30] In his early campaigns in the West, Halleck had ordered no less than 100 rounds per man and, around Knoxville in 1864, 35th Massachusetts received the same number.[31] 'One of Jackson's foot cavalry' tells us that on 'special occasions' his load would be 80 rounds,[32] although the general impression one gathers from much of the literature is that the Confederates were generally less likely to make extra issues than were the Federals. The South had far less ammunition to spare, they had less good-quality powder, and on some occasions they took specific measures to forbid profligate firing.[33] (Nevertheless, at least one Confederate was captured with all of 200 rounds on his person.[34]

The trouble with these extra loads of ammunition was that they were often issued at just the same time as extra rations of food. Thus the soldier was doubly weighed down at precisely the moment at which he needed exceptional lightness for exceptional manoeuvres and exertions. It was common to find him taking a number of short cuts to avoid this dilemma, but in the end none of them was truly satisfactory. If he ate all his rations at a single sitting he would go hungry later. If he dumped his ammunition surreptitiously he would find himself disarmed at the critical moment of combat. And if he deposited his pack or blanket roll with a sentry for the duration of the battle he risked returning to find it plundered – or never finding it again at all. There was also a slightly less dramatic practical problem when it came to finding a place to physically carry the extra rounds. With a full pack and a full ammunition pouch there were precious few places left vacant in the soldier's clothing for this ammunition, and certainly none which were truly convenient or comfortable. The pockets were obvious candidates, as were knapsacks, but in neither case was the end result a particularly happy soldier:

During the day there were issued to each of us 80 rounds of rifle ammunition. As our cartridge boxes would only hold 40, the extra had to be put in our knapsacks. We were issued eight days' field rations consisting of hardtack, coffee, pork and sugar. Since our haversacks only had capacity for three days, the extra rations also had to be stowed in our knapsacks. We were notified that under no circumstances were we to part with either our rations or ammunition, except when their use became necessary. There would be no further issue of food or ammunition for eight days, unless ammunition became exhausted in action. By close packing we succeeded in getting the additional rations and ammunition in our knapsacks which after they were strapped in the usual way presented a very inflated look. When we slung them on our shoulders they were heavy.[35]

Obviously the interests of higher commanders lay in giving their regiments the maximum possible firepower, yet they had to balance this against the likelihood of soldiers jettisoning ammunition if they felt they were being overloaded. It was a rather delicate distinction to draw, and we find that in practice there were all too many regiments which ran out of ammunition and gave up the fight before a decision had been reached in any other way. Perhaps we notice this particularly because Civil War records have been better preserved than Napoleonic ones, but the fact remains that there do seem to be proportionately rather more well-documented cases of ammunition shortages in firefights during the later conflict than in the earlier.

In his book on the battle of Seven Pines the Confederate New Yorker, General Gustavus W. Smith, accuses General Longstreet of fabricating several tales of 'running out of ammunition' simply to excuse his failure to attain his tactical objectives.[36] Whatever the rights and wrongs of this particular case (which are very unclear indeed), we at least have to agree that similar excuses may well often have been used to cover the tracks of regiments which were less than enthusiastic for the fray. Either they might have thrown away their surplus ammunition in the approach to combat or they might have given up the struggle before every recourse (such as organising reliefs or calling up a resupply wagon) had been exhausted. Not all of the recorded cases of 'running out of ammunition' in combat should therefore be accepted automatically at their face value.[37]

Yet on the other side of the coin we find that the arithmetic does work powerfully against high volumes of fire being sustained for very long periods. If each man carries forty rounds and fires one round per minute – a relatively slow rate in theory, although far from slow in practice[38] – it is clear that the regiment will be forced to cease fire after only forty minutes. This is not to say that firefights were necessarily short, since we shall see later that they often dragged on for hours at a time. What it does suggest, however, is that prolonged firefights must have been run at very low rates of fire – perhaps only one round per man per five or even ten minutes. The alternative was to burn through the entire ammunition stock very rapidly, leaving no options but retreat or a desperate charge with cold steel.

It is to the credit of many regiments that in the event they chose the second of these two options. At Gettysburg 20th Maine made a bayonet charge when its ammunition had all gone, as did 14th Indiana at Green Brier on 2 October 1861.[39] At Second Manassas the Confederate defenders of the railroad embankment took to throwing rocks in lieu of bullets,[40] and so on. It is equally to the credit of other regiments that they succeeded in scraping together some supplementary ammunition before it was too late. Some of Gustavus W. Smith's own units at Seven Pines organised a series of reliefs whereby different troops took their turn in the line some five times over, to maintain the volume of fire upon the enemy.[41]

Few Civil War diarists are very specific about precisely how many rounds they fired in combat, but when they are it is usually because the figure is felt to be sufficiently exceptional to be worth recording, or even boasting about. The physical labour involved in firing even forty rounds was itself considerable, to which we must add the bruising of the shoulder by the recoil, the caking of the face with powder and the inhalation of acrid fumes:

At Peach Tree Creek I fired seventy rounds of ammunition, some of the men said they had fired one hundred. At the end of the action we presented a strange appearance, smoke and powder stains had covered our faces and made them look as blue as indigo. The day had been hot and we were as wet as if we had been in the water.[42]

Compare this experience with that of Sam Watkins after the battle of Kenesaw Mountain:

after undressing [I] found my arm all battered and bruised and bloodshot from my wrist to my shoulder, and as sore as a blister. I had shot one hundred and twenty times that day. My gun became so hot that frequently the powder would flash before I could ram home the ball, and I had frequently to exchange my gun for that of a dead colleague.[43]

We nevertheless do sometimes hear of even higher numbers of rounds being fired in a single day, although how far such tales can be trusted is open to debate. Private Milledge of 66th Ohio claimed to have fired two hundred rounds at Gettysburg,[44] for example, while G. Norton Galloway of 95th Pennsylvania said that he and several men of that regiment each fired three or four hundred rounds of ammunition *or more* at Spotsylvania.[45]

The cases cited above are likely to have been exceptional, since the total ammunition consumption in battles appears to have averaged a far lower figure overall. For Gettysburg we have a Confederate Ordnance estimate that each man fired an average of 25–26 rounds,[46] although the precise basis for these figures is not clear. They seem to refer to the rounds presumed fired during the whole week in which the battle fell, by all 75,000 Rebel troops in the general area. The figures may in fact reflect the correct order of magnitude for ammunition expended in the battle itself, but there is no certainty. If they *are* accurate, we can set them beside Union casualties of some 23,000 men to arrive at a figure of 81 shots fired to inflict each casualty, or maybe nearer to 100 infantry shots per casualty if we also count in the contribution of the artillery.

Rather better figures are available – as in so many other aspects of the war – for the Union forces. We find that Meade's 90,000 men were issued a total of 5,400,000 rounds at Gettysburg, giving an average of 60 rounds per man, although not all of these may actually have been fired. Rather higher figures are reported in specific formations; thus Geary's brigade supposedly fired 75 shots per man and I Corps fired no less than 86. As a result of all these fusillades Lee's army lost some 20,000 men.[47]

If we estimate the overall average actually fired as lower than the number of rounds issued, we can guess that the average Union soldier really fired only 40 rounds in the three days of the action. These calculations give a notional 180 rounds fired for every casualty inflicted by the Federals, although that is without counting the artillery's contribution. A fair guess would be that each casualty caused by infantry fire required the expenditure of 200 rounds. This is higher than the rather unreliable figures for the Confederate side, but consistent with the order of magnitude recorded for the Napoleonic Wars.[48]

The sad fact is that the weapons used at Gettysburg fell a long way short of perfection. The sequence of actions required to load and fire a rifle musket was still extremely complex and difficult to follow correctly in the heat of combat (see Table 3.5). The drill manual included seventeen separate movements for each shot fired, which compares very unfavourably with the movements needed to fire even the simplest breechloader. Even with well drilled men it was likely that many rifle muskets would become inoperative in combat as a result of errors or fumbling in the loading sequence. It is partly for this reason that we often hear of cases in which musketry was ineffective, for example a bloodless skirmish behind Vicksburg on the night of 25–26 June 1863:

It seems strange however that a company of men can fire volley after volley at a like number of men at not over a distance of fifteen steps and not cause a single casualty. Yet such was the facts in this instance.[49]

A major part of this ineffectiveness must have derived from poor aiming, and another important part must have come from misfires. These are mentioned quite often in the literature and sometimes occurred on a grandiose scale. At Shepardstown, 20 September 1862, we read that the 'Corn Exchange Regiment' suffered no less than a 25 per cent rate of misfires.[50] Nevertheless, another important proportion of the rifle's ineffectiveness must certainly have derived from faulty loading.

TABLE 3.5   DRILL MOVEMENTS TO LOAD AND FIRE ONE ROUND [51]

1836 smoothbores	18
Percussion rifle muskets	17
Early breechloaders	6
Spencer repeaters	4
Henry repeaters	2

When rifles were fired repeatedly in quick succession they overheated and might detonate before loading was complete.[52] Alternatively, a soldier in too much of a hurry might forget to extract his ramrod, and fire it at the enemy along with his bullet. At Spotsylvania Major Ellis of 49th New York was wounded in the arm and body by a ramrod fired in this way.[53] A few days later the men of 13th Massachusetts must also have seen a ramrod being fired, for they made it into a game to be repeated on several occasions. Apparently it was the curious whizzing noise of this strange projectile that gave it its attraction.[54]

An often quoted set of statistics from Gettysburg has it that the Union forces salvaged 27,574 'muskets' after the battle, of which 24,000 were loaded, including 12,000 loaded twice, 6,000 loaded between three and ten times, one with twenty-three charges and one with twenty-two balls and sixty-six buckshot. Some had six balls and only one charge of powder; others had six unopened cartridges. Others again had the ball behind the powder instead of the other way round.[55]

It is open to doubt whether twenty-three full cartridges could in fact be physically squeezed into the barrel of a Civil War rifle,[56] and still more dubious that the proportion of misloaded weapons in the sample (some 45 per cent) actually reflects the proportion in the whole of the two armies during combat. It is most likely that many of the guns salvaged by the Union forces after Gettysburg were discarded by their users precisely *because* they had become unusable, hence the figure of 12,000 should be seen as a proportion of the total muskets in the battle rather than of the total salvaged. That suggests that perhaps 9 per cent of all muskets were misloaded – a less dramatic figure, but nevertheless still very significant. If we add the unknown total of misloaded muskets which were either salvaged by the Confederates or retained by their original owners, we are forced back to the conclusion that a very high proportion of infantry weapons must indeed have become inoperative in combat due to faulty handling.

RIFLE TRAINING

In the stress and emotion of battle the soldier could load and fire his rifle correctly only if he had long experience in its use. This experience might consist of familiarity with firearms in civilian life, before the individual had joined the army; it might consist of 'dry' drilling in the seventeen movements required by the manual; or it might consist of live firing practice, either on the target range or on previous battlefields. Just how much of each of these was available in any given case would clearly vary enormously from one man to the next and from one week to another, but it may be useful for us to attempt some broad generalisations here.

It has often been alleged that Americans were more accustomed than Europeans to handling small arms in civilian life. The farmer kept his shotgun or hunting rifle, and the Western pioneer had his pistol for personal protection. In the cities of the north-east there was perhaps a more 'European' lack of firearms, but even in states like New York or Massachusetts we find that a large proportion of the men enlisted for the Civil War actually came from the countryside rather than from the town. When we look down south into Dixie, of course, we find an ever stronger association in the popular mind between the white citizen and his gun. The 'typical' Confederate is generally imagined to have been born and bred with a rifle by his side, or if not a rifle then at least a duelling pistol.

This image may be severely questioned, however, in view of the heavily political motives which lie behind it. Michael Adams has well demonstrated how opponents of slavery wished to portray all members of a slave-owning society as personally violent people, and hence deliberately built up the image of the gun-toting Southron pretty much independently of the facts.[57] Besides, even if we do accept that knowledge of firearms was somewhat more widespread in America than in Europe in the 1860s we should beware of exaggerating its implications for behaviour in battle. Whether he is a French peasant chasing starlings or a Blue Ridge hillbilly going after racoon, the huntsman who loads carefully and then stalks his inoffensive prey is surely in a very different state of mind from the soldier who has to fire off forty rounds in double-quick time against an enemy regiment which is busy returning the compliment. The assumptions of the close-order firefight, in other words, are surely located in a quite different universe from the genteel expectations of game shooting.

Few Civil War regiments seem to have experienced any sort of convincing 'battle inoculation' for live firing in close order, however often they may have run through the drills in dumb show without ammunition. There does appear to have been a serious lack of target practice in the armies of both sides, and we find that when it did occur most diarists regarded it as a highly exceptional event. The Union troops in the East generally seem to have done a bit of range shooting in the early spring of 1864 in preparation for the Wilderness battles, but apart from that the impression is that range days were almost unknown on any regularly organised basis.[58] It is far more common to read of massed inter-brigade snowball fights than of live firing, even within a single regiment. We may speculate, perhaps, that short-range snowball drill gave a better imitation of real battle than any exercise which involved aiming rifles for long-range fire. In season the snowball was also available in unlimited quantities, which is more than either side could say of live ammunition.[59]

If we look at the case of 24th Michigan, we find that it was sent to the front within a very few weeks of its formation in July 1862, and in its only recorded target practice during that time three men were wounded and one died of a heart attack. In these circumstances we may be forgiven for imagining that live fire was almost as dangerous to the men who were delivering it as it was to the enemy.[60] At all events the regiment's next target practice came some four months later, but was not followed through. After this we learn of a resumption over a year later – long after some 80 per cent casualties had been sustained at Gettysburg – and the regimental historian assures us it was the first time this form of training had been stressed.[61]

In the history of 35th Massachusetts, which was formed in mid-1862, we find plenty of references to drill and tactical training, but none to target practice. There was apparently a feeling in many camps that if such firing were allowed it might start a panic among other troops hearing the volleys, who might imagine there was a real battle in progress. Nevertheless, the lack of target practice does seem to have been regretted by members of the regiment: '... although the Government had adopted rifled arms, target practice was never encouraged; men learned the use of their weapons in battle or by stealth.'[62] In a sister regiment, 13th Massachusetts, there was a similar story. The regiment was formed in August 1861, it spent a great deal of time and effort on drilling, but held its first target practice (organised at brigade level) only in the spring of 1864. In the intervening period it had fought in half a dozen battles, but its nearest approach to formal target practice had been an exercise in estimating ranges up to five hundred yards.[63]

The estimation of range was a matter of vital importance to any long-range firing with rifle muskets, since their low velocity meant that the 'point blank' zone was relatively short. Beyond a hundred yards or so the shots would have to arc high in the air to come plunging down on their targets at a steep angle. If the sights on the rifle were badly set, this angle would be wrong and the shots would miss. Hence accurate firing at long ranges was a particularly delicate operation.[64]

There is little in the literature to suggest that the average Civil War infantry regiment even began to judge distances or set sights accurately for battle. On the contrary, there are many references to officers telling their men simply to aim low in order to counteract the natural inclination of a flinching musketeer to shoot high. The first article of the Confederate General Hindman's orders to his men at Prairie Grove, dated 4 December 1862, urges them to

Never fire because your comrades do nor because the enemy does nor because you see the enemy nor for the sake of firing rapidly. Always wait until you are certainly within range of your gun, then single out your man, take deliberate aim as low as the Knee & fire.[65]

These sound counsels are surely designed to produce effective fire at short rather than long ranges, and are based upon a deep scepticism that the average soldier will remain calm and responsive to orders when the enemy comes into view. It seems that General Hindman, at least, expected his men to waste their fire at long range unless they were carefully supervised by their commanders.

Admittedly in this particular case the troops who were being asked to aim low were probably armed with smoothbore weapons. They could not have fired accurately at long ranges even if they had been inclined to do so. Yet even with good rifles the same advice would still have been appropriate, since the real problem with Civil War infantry fire always lay in the most rudimentary aspects of maintaining the men's concentration on their job. In combat, fire would usually be delivered from a line two ranks deep, with the soldiers in each rank almost touching each other's elbows. They would naturally jostle and shove each other as they drew their ramrods and pushed home their cartridges. Anyone in the second rank would have to lean forward to fire through the space between two men in the front rank, who would receive a flash and a cloud of smoke in their eyes and a numbing explosion at the level of their ears. The target would be obscured and the orders of officers attempting to direct the fire would be drowned in the din. Effective fire control would become impossible. Each man would be thrown back on his own ability to plug on, loading and firing doggedly within his own cocoon of emotions. It would be little wonder if the final result fell far short of an efficient military machine.

The Union soldier James Tinkham reports the scene as it struck him at First Manassas: 'The men were a good deal excited. Our rear rank had singed the hair of the front rank, who were more afraid of them than of the Rebels.'[66] Another account, this time from 1st South Carolina, outlines the way these pressures tended to make the soldiers shake out from their regulation close-order formation into something a little less rigid:

A battle is entered into mostly in as good order and with as close a drill front as the nature of the ground will permit, but at the first 'pop! pop!' of the rifles there comes a sudden loosening of the ranks, a freeing of selves from the impediment of contact, and every man goes to fighting on his own hook; firing as, and when he likes, and reloading as fast as he fires. He takes shelter wherever he can find it, so he does not get too far away from his company, and his officers will call his attention to this should he move too far. He may stand up, he may kneel down, he may lie down, it is all right – though mostly the men keep standing, except when silent under fire, then they lie down ... A battle is too busy a time, and too absorbing, to admit of a good deal of talk. Still you will hear such remarks and questions as: 'How many cartridges you got?' 'My gun's getting mighty dirty.' 'What's become of Jones?' 'Looky here, Butler, mind how you shoot; that ball didn't miss my head two inches' – 'Just keep cool, will you; I've got better sense than to shoot anybody' – 'Well, I don't like your standing so close behind me, nohow.' 'I say, look at Lieutenant Byson behind that tree.' 'Purty rough fight, ain't it Cap'n?' 'Cap'n, don't you think we better move up a little, just along that knoll?' All this is mixed and mingled with fearful yells, and maybe curses too, at the enemy.[67]

The Civil War rifle musket was unquestionably a far superior instrument to the Napoleonic smoothbore flintlock which it gradually came to replace, and in theory it could fire accurately at more than three times the range. Yet the rifle musket was not available to all regiments until late in the war, nor was it always as mechanically efficient as could have been wished. There was no great superabundance of ammunition, least of all in the South, nor was there sufficient industrial capacity for a massive conversion to breechloaders, even in the North. The individual soldier was usually limited to a rather meagre supply of cartridges, allowing heavy firing to be sustained by a regiment for only a relatively brief period. An almost total lack of target practice meant that many rifles were misloaded in combat and that the finer points of long-range accuracy were neglected or ignored. The close-order drill of the day also meant that the soldier in battle was subjected to a barrage of sights, sounds and emotions which must have distracted him powerfully from his task. Even with these wonderful new weapons, in fact, it remains doubtful that a genuine revolution in firepower had actually occurred. It might be more reasonable to talk of some minor improvements over Napoleonic performance, but little more than that.

4

Drill

But he instantly saw that it would be impossible for him to escape from the regiment. It inclosed him. And there were iron laws of tradition and law on four sides. He was in a moving box.

The Red Badge of Courage, p.21

We have seen that the Civil War soldier usually lacked an adequate basis of target practice before he went into action, particularly in the type of sustained close-order firing which combat often required. We may therefore say that his training was defective in this respect, just as it was defective in various other skills which are given prominence in twentieth-century military curricula. There was no 'battle inoculation' for coming to terms with casualties or with near misses from enemy fire. There was little attempt to exercise in broken terrain or at night, and there were no manuals of tactics in the modern sense of the term. The recruit's acclimatisation consisted of just two elements, neither of which was of very direct relevance to what he would be expected to do once the battle had been joined. These two elements were, first, the experience of survival within the regimental community as it marched, ate and slept; and, second, an almost endless experience of drill.

THE REGIMENTAL MATRIX

The theoretical strength of a Civil War infantry regiment was about 1,000 officers and men, arranged in ten companies plus a headquarters and (for the first half of the war at least) a band.[1] There were major differences at various times and places, including a provision for extra sub-units to be added if additional recruits were available. Thus 66th Georgia counted 1,500 men at its formation in 1864, arranged in one 'regiment' of ten companies and one independent 'battalion' of three.[2] 36th Illinois started in July 1861 with 1,151 men in ten companies of infantry and two of cavalry,[3] while 14th Indiana at the same time was using 1,134 men simply to fill out its ten infantry companies.[4] From these examples we can see that the regiments were not universally standardised and did frequently begin their campaigning with more than the theoretical establishment of troops.

What was an almost universal rule, however, was that before many months had passed the regiment would be very seriously reduced in numbers, even without counting any battle casualties:

Something like one fifth of the men who enlist are not tough enough or brave enough to be soldiers. A regiment reaches its station a thousand strong; but in six months it can only muster six or seven hundred men for marching and fighting duty; the rest have vanished in various ways. Some have died of hardship, or disease, or nostalgia; as many more have been discharged for physical disability; others are absent sick, or have got furloughs by shamming sickness; others are on special duty as bakers, hospital nurses, wagoners, quartermasters' drudges &c.; a few are working out sentences of court martial.[5]

The process of acclimatisation to camp life would also involve the wholesale jettisoning of impedimenta, as it was found that overanxious greenhorns had encumbered themselves with far more arms, equipment, bibles and keepsakes than they could conveniently carry on the march. Sam Watkins of 1st Tennessee described his first tiring day's march as follows:

From the foot to the top of the mountain the soldiers lined the road, broken down and exhausted. First one blanket was thrown away, and then another; now and then a good pair of pants, old boots and shoes, Sunday hats, pistols and Bowie knives strewed the road. Old bottles and jugs and various sundry articles were lying pell-mell everywhere.[6]

This description comes from the Confederate army, which – in contrast to the Union – could never lay hands upon as much equipment as it needed. The wastefulness may thus be seen as somewhat surprising. Yet it is not at all untypical of what was happening to the men of both sides as they shook off civilian misconceptions, toned up their muscles to the routine of marching, and reduced their personal baggage to the minimum necessary for life on campaign.

After a while only the truly fit would be left,[7] although even they would continue to be reduced in numbers by the various hazards of war. The average fighting strength for a regiment was perhaps 400, with the North managing to field a few dozen more than the average and the South a few dozen less.[8] Within these round numbers there were of course enormous variations for specific cases, and many were the regiments that fell so far below a reasonable figure that they disbanded some of their companies or even went into battle at effectively company strength. Thus 13th Massachusetts on 11 May 1864 mustered a mere 107 men; 20th Maine in late 1863 had 80, and 17th Virginia at Antietam could put but 55 soldiers into the line.[9] We might certainly suggest that the average strength for a well seasoned regiment which had seen its fair share of action was perhaps a half of the figure of 'around 400' which represents the average for the armies as a whole.

Such small units had been rare in the Napoleonic Wars, yet in the Civil

War they were accepted as a matter of course. It was found that their relative lightness was actually a tactical advantage, and late nineteenth-century commentators noted that they were consistent, albeit accidentally, with the general trend of the age.[10] They had more in common with the German practice of fielding large independent companies than with the French preference for keeping several small ones together in a battalion. When stripped down for action a Civil War regiment of two or three hundred veteran troops made a very handy and manoeuvrable fighting unit. Although it might not run to a commissioned officer present with each of its ten companies, it could nevertheless be commanded easily from the centre, by voice and personal example.

Much of a regiment's resilience in battle depended on the length of time its men had been campaigning, and especially the length of time they had been living together as a unified team. Institutional continuity was of great importance for morale, particularly at the lower levels of command. Thus it scarcely seemed to matter to members of Burnside's IX Corps when they were shuffled from the Virginia theatre to the Carolinas, then to Kentucky, then on to Vicksburg and back to Kentucky and Virginia again. They remained the IX Corps regardless of who commanded the 'section' or 'army' to which they were attached.[11] Equally, a division could change its corps without too much friction, and a brigade might even change its division – although that might sometimes cause problems. When a regiment changed its brigade, however, there was an almost inevitable period of suspicion to be endured before it would be fully accepted, and this could sometimes spill over into failures of co-operation and co-ordination on the battlefield.[12]

Still more extreme were the problems encountered when company-sized groups were transplanted into regiments of which they had not previously been a part. Thus a proud and patriotic remnant of a unit disbanded after completion of its three years' service might opt for re-enlistment only to find itself a resented minority in a cocky young outfit which did not even come from the same state. Alternatively, a large group of raw recruits might suddenly arrive in a veteran regiment, to be greeted by the jeers of old soldiers who 'knew it all' already. Such greenhorns would quickly be assimilated if they were men of character and good will, and particularly if they were spread out among the four-man buddy groups rather than kept together in a single mass. But if there were any obvious differences of motivation, background, nationality or even county of origin, then bad blood and delays in acceptance by the old hands might well be the result.

There are many examples of poor assimilation into regiments, but perhaps the most spectacular is the case of 13th Massachusetts on 14 August 1863. On that day – incidentally at the height of the Northern draft crisis – the regiment took receipt of 186 new recruits. Of the total batch 50 per cent were found to be masquerading under assumed names. Forty of the newcomers deserted on the first night, and a further seventy-five on subsequent nights. Twenty-six had to be transferred to the Navy and six more were discharged sick. The effective benefit to the regiment was therefore but thirty-nine men, or just over 20 per cent of the new arrivals. This contrasts starkly with a battle loss suffered by the regiment at Gettysburg, six weeks earlier, of some 185 killed, wounded and prisoner. It may be that the regiment was still too much affected by these heavy casualties to be in a mood to welcome new blood, but it seems more likely that the problem arose from the nature of the new recruits themselves. They were renegades from all walks of life, including some Confederate deserters and a rich variety of New Brunswick plumbers, Scottish and English seamen, Belgian shoemakers, Prussian machinists and Irish labourers. A more toothsome assemblage of foreigners and bounty-jumpers would be difficult to find anywhere.[13]

The story of 13th Massachusetts's Gettysburg replacements does at least give the lie to the popular myth that the Union Army failed to top up its regiments in the field, unless they happened to come from Wisconsin.[14] Even if that particular state did enjoy a better system of individual replacements than any other, it is quite erroneous to suggest that the regiments of the others were allowed to wither away without receiving any replacements at all. Both 18th Missouri and 20th Maine, for example, received three major new drafts in the space of three years – and this was not untypical.[15] The problem lay rather with the unmanageable size and unpredictable timing of reinforcement drafts, coupled with the distracting personal effort in recruiting and administration which they demanded from officers serving in the regiment. The system was certainly clumsy and unpopular with everyone from General Sherman downwards – but that is not quite the same thing as asserting that no system of any kind existed.

We should also beware of following the common assumption that the Confederacy had a sophisticated system for drafting individual replacements as and when they were needed. This was not the case, as is amply testified by the relatively low fighting strength of Rebel units and by the fact that new regiments continued to be formed right up to the end of the war, thereby diverting newly recruited manpower away from the old units. Nevertheless, it is true to say that more men were rotated through each Confederate regiment, on average, than through each Union one. Table 4.1 gives some indication of the figures.

Despite the many rough edges on these statistics, we can at least feel safe in stating that each Confederate regiment on average saw more action than each Union one, and was pulled out of the line for rear echelon work less often. Because there were so relatively few Confederate regiments each one of them, on average, had to bear a greater share of the fighting

TABLE 4.1   MOBILISATION OF REGIMENTS IN THE CIVIL WAR[16]

	Confederate	Union
Mobilisable population	5,500,000	22,000,000
Number recruited at one time or another	950,000	2,100,000
Total infantry (after deduction of 20% for non-infantry units)	740,000	1,680,000
Number of infantry regiments fully formed	500	1,320
Average men passing through each regiment	1,480	1,272
Approximate battle losses (killed and wounded)	340,000	375,000
Average killed and wounded per regiment	680	284

Note:

All figures are 'guesstimates' intended to indicate an order of magnitude only. They are not, for example, based on any certainty of the numbers of infantry regiments raised primarily for combat (as opposed to merely labouring) duties. Nor are the many auxiliaries/irregulars, who were not formally enlisted, necessarily included. Probably very many more people than indicated here were combat participants in this war.

than its Union counterpart. By the same token each of them received rather more replacements — perhaps 450 in the course of the war as against some 250 for the average Union regiment – to make up its correspondingly greater battle losses.

In terms of combat quality we may certainly assume that the Confederate regiments, being generally more experienced, ought to have enjoyed a certain advantage over their opponents. Admittedly, many of them must have been well on their way down from 'veteran' into 'old lag' status by the summer of 1863, although that process could still be balanced to some extent by fresh units making the upward leap from 'greenhorn' to 'veteran'. A year later the picture was not quite so bright, and we can almost talk of whole armies of old lags by then.

Length of service was not, of course, the only determining factor in a regiment's quality, since leadership also had an important part to play. One statistical pointer which may be of relevance here is that in a sample of 48 battles studied by Thomas Livermore the Union officers present numbered between 4 and 7 per cent of their total forces, whereas the figure for the Confederates was between $6\frac{1}{2}$ and 11 per cent.[17] Thus the Confederates seem to have enjoyed a higher ratio of officers to men, which is normally taken as a valuable indicator of higher combat quality in an army as a whole.

When it comes to the quality of the individual officers, by contrast, the picture is less clear. Much ink has been spilled in discussion of the respective merits of academy-trained officers as against 'hostilities only' men drawn from civilian professions. The practice of electing officers has also proved controversial, for on one hand it can be seen as a necessary safeguard against political interference by state governors, while on the other hand election is itself a highly political method of appointment. The men it places in high office may not necessarily possess high professional qualifications, and there can be little doubt that in the Civil War many officers on both sides were actually unfit for command.[18] Nevertheless, the circumstances in which these armies were raised were themselves highly exceptional, and it would have been astonishing indeed if a large number of unsatisfactory officers had *not* been appointed. This was particularly true of the first two years, which were also the years in which elections flourished and then declined.

We must always remember that the pre-war United States Army had numbered a mere 16,000 men. From its pool of serving or retired officers only 923 came forward to fight for the Union in the Civil War, and 369 for the Confederacy.[19] Maybe that number could be doubled if one includes men educated at a private military school, or with military experience as a volunteer in the Mexican or Indian wars, but it still leaves less than 3,000 trained officers available to command some 3,050,000 troops, or little more than one officer to each regiment. This contrasts with a theoretical requirement for some 39 commissioned officers in each infantry regiment, plus an even higher proportion in the technical arms and the staff. In these circumstances it was inevitable that more than 98 per cent of Civil War officers would have to learn their job during the war itself. The amazing thing is therefore surely not that some of them failed to rise to the challenge, but that so many of them succeeded in such handsome style.

Between North and South there was perhaps rather less difference in the quality of leadership than the popular stereotype of Yankee incompetence versus Southern brilliance would lead us to suppose. Admittedly, the rapid expansion at the start of the war – by its very scale – created more problems for the Union, and the complaints against poor appointments were louder in the North than in the South. But for that very reason it was the Union that abandoned the election of officers first,[20] and the Confederacy that exhausted its reserves of promising officer material more quickly. In leadership talent, as in so many other things, it was ultimately the North that proved to have the deeper pool of resources.

None of this, however, completely absolves the Civil War governments from the charge that they failed to provide the infrastructure necessary for the efficient creation of regiments. Unlike the large European conscript armies, the pre-war United States Army had not been designed to act as a *cadre* for wartime expansion, nor in the event did it do so. The vast majority of Civil War regiments therefore had to be trained and acclimatised to camp life by officers and NCOs who were themselves new to the game. Contrary to European practice, wartime expansion had to be conducted without the benefit of long-service professionals, without the benefit of pre-existing regimental traditions and institutions, without territorial recruiting depots, and without the doctrinal analysis and guidance which might have been provided by a true general staff. If senior Prussian commanders such as von Moltke looked rather askance at such an army, they had some excellent reasons for doing so.

In the United States before the war the opinion of the professional experts – men like Halleck, McClellan and Winfield Scott – had also been realistically lukewarm about what could be achieved with such a hastily improvised force. Once hostilities had started, furthermore, these professionals felt duty-bound to tone down the uninformed optimism of politicians who believed that any American army – improvised or not – could win the war in a day. Far from being a widespread incitement to recklessness, we can see the influence of West Point as a warning in favour of caution – a warning which was ultimately swept aside because of West Point's own failure to provide adequate numbers of trained officers when war arrived. The pre-war army may be blamed for not preparing for a conflict on the scale that was required – but it can scarcely be blamed for underestimating the difficulties once the shooting had begun.

Amid the general relief that the armies actually fought as well as they did – or even that they fought at all – some of the valid criticisms of the professionals have tended to be lost from view. For example, Sherman, who in many ways is seen as the opposite of a European-style disciplinarian, was quite clear that discipline was a major problem in the First Manassas campaign:

The march demonstrated the general laxity of discipline; for with all my personal efforts I could not prevent the men from straggling for water, blackberries or anything on the way they fancied ... We had good organisation, good men, but no cohesion, no respect for authority, no real knowledge of war. Both armies were fairly defeated, and whichever had stood fast the other would have run.[21]

Admittedly efforts were made to tighten discipline after First Manassas, but we have an unmistakable impression from the literature that they never quite achieved the full effect desired:

The weak point of the volunteer service was that the soldier detested taking orders and the officer shrank from commanding. The soldier would comply with a reasonable order but he did so because it was reasonable, not because it was an order. If seemingly unfair or trifling orders were issued, the officers found that they had not soldiers but free and independent citizens to deal with ...

... often there was an easy familiarity between officers and men which made rigid discipline impossible. Too often both officers and men came from the same neighbourhood, to which they all intended to return. They had known each other and each others' peculiarities all their lives. The privates could not bring themselves to call by anything but his first name, or to salute with decorum the village lawyer, undertaker, or livery-stable flunkey, even if he were now a captain or a lieutenant.[22]

In the Eastern theatre – both in the 'Bandbox Army of the Potomac' and among 'Lee's well disciplined soldiers'[23] – the drive to build discipline was eventually more successful than in the more independently minded Western armies. Perhaps this was because there was a more European flavour to Eastern culture, or because the political pressures were stronger there, or because there were more unwilling conscripts in need of close supervision. Maybe it was simply that the Eastern city-dwellers were naturally more disciplined in their civilian lives than frontiersmen who lived in log cabins or shanty towns. Whatever the reason, it is still probably fair to say that by the end of the war the armies in the East had reached an *almost* European standard, while those in the West retained – and even positively gloried in – their own inimitable character.

One veteran who returned to the Army of the Potomac in early 1864 after a year in the West reports that 'We remarked among the new troops a harsher discipline than prevailed in the army of 1862'.[24] For him, as for so many others, the contrast between East and West and early-war and late-war was unmistakable. An experienced *cadre* of officers and NCOs was at last beginning to make its mark upon the Eastern armies, and the naiveties of the first two years had passed away – although even in the East the full rigours of European discipline were never quite attained. This is not to say that the average Civil War army, whether in the East or in the West, was particularly inefficient or slack. It wasn't. But, as the French commentator de Chanal explained, it was certainly 'different':

At the core and in all that is essential, its discipline is as good as, if not better than, that of the European armies; but it has not the external marks, and an observer who merely passes through an American army may thus be deceived.[25]

Captain De Forest made essentially the same point from his own experiences as a company officer:

The soldiers are as obedient and quiet as sheep. But they don't touch their caps when they meet an officer; they don't salute promptly and stylishly when on guard; in short they are deficient in soldierly etiquette.[26]

De Chanal and De Forest are thus agreed that it was only in the outward appearance and etiquette of military smartness that the improvised American armies fell down – and these aspects must indeed have held but trifling importance in camp or on the march. When we come to the way battles were fought, however, we find that appearances and etiquette may not have been quite so trifling after all. The awkward fact, for anyone who would dismiss European military culture as irrelevant to the Civil War battlefield, is that the very basis and fundamental structure of tactics came out of the European drill manuals. If the American armies failed to master these drills, in other words, there would be a sense in which they failed to master the whole art of fighting.

THE DRILL BOOKS OF THE 1860s

The first full American drill manual had been issued in 1779, under the auspices of the German mercenary adventurer von Steuben. As was only fitting, it was based upon the drills of the leading military power of the day – the Prussia of Frederick the Great. As time went on, however, the infant Republic grew closer to France than to Germany, until by 1812 it was the French drill book that was being reproduced for American use. By the time of Waterloo and the wave of nostalgic Bonapartism which followed it, almost every corner of the American military stage was filled with French ideas and French models, ranging from the coastal fortifications of Bertrand to the strategic geometry of Jomini. When the French 1791 drill book was revised by Brenier and Curial in 1831 it was therefore faithfully translated almost at once by the foremost American General, Winfield Scott, and issued as the *Infantry Tactics or Rules for the Exercise and Manoeuvres of the United States Infantry* in 1835.[27]

Scott's *Tactics* was a three-volume work. The first volume deals with the training of the individual soldier and his movements within a company. It envisages a line three ranks deep, although the option of using two ranks is still left open. This line is intended to march forward at 90 or 110 paces to the minute, or fire three rounds of musketry per man per minute. Sound practical advice is offered on such topics as the need to keep distances carefully when marching, or to beware of double-loaded muskets when there have been misfires.[28] The drum signals for ceasing fire are especially stressed as 'of great importance with troops, not veteran, in the face of the enemy'.[29]

The second volume deals with the movements of a regiment (or 'battalion') of ten companies, and with the skirmishing procedures for light infantry. Much space is taken up with explanations of five different ways for deploying the battalion line into a column of companies, and then deploying these back into line. There are also five different ways for the battalion to deliver a volley – by company, by wing, by file, etc – a section on forming squares against cavalry and a great deal on how the colour party at the centre of the battalion should be used to give direction and orientation to the whole. Although much of this looks pedantic and impractical to the outsider it was based on thirty-five years of French experience with their 1791 drill book, culminating in a series of special trials, in 1826–7; so presumably it all had greater value than meets the eye.

The instructions for skirmishers are intended to be less pedantic and more general in nature, although some sort of alignment is still supposed to be kept in the skirmish line. The main rules are that every company deployed to skirmish shall always keep back one-third of its strength as a formed reserve, while within the skirmish line itself the men work in pairs, alternating their fire in order to have at least one rifle loaded at any time.

Although the flank companies of the battalion are designated as specialist skirmishers and are expected to be the only troops with rifled weapons, it is also fully accepted that any or all of the other companies may from time to time join in the skirmishing. In Scott's tactics there may thus be circumstances in which the entire battalion front consists of dispersed individuals lying or kneeling behind cover rather than standing together in close order. Such a formation was by no means new to the military art in 1835, but Scott's drill book did perhaps help to make it a little more commonplace than it had been in the past.[30]

The third volume discusses the manoeuvres of a line of eight battalions, forming an 'army corps' of two divisions, each of two brigades. In the Civil War brigades would actually consist of anything from two to eight (or even more) regiments;[31] divisions anything from two to five brigades, while army corps could be still more variable than that. Nevertheless, Scott's schematic layouts show us how he and his French mentors felt any force of multiple battalions ought to manoeuvre in line or column, including changing front and negotiating defiles. This section is thus a handbook for brigadiers and above, with somewhat less everyday application than the earlier volumes intended for use within a battalion or a company. Indeed, we may speculate that while company officers may often have used the full procedure 'according to Scott' for their manoeuvres on the battlefield, brigade and higher commanders may just as frequently have departed from them and transmitted their orders informally.

Scott's manual was successfully used by the US invaders of Mexico in 1846, and it continued to be reissued at least until 1861. Its general structure remained at the heart of all the Civil War drill manuals. In 1855, however, a fuller and more modern version of the first two volumes appeared, in the shape of Lt-Col. William J. Hardee's *Rifle and Light Infantry Tactics for the Exercise and Manoeuvres of Troops when acting as Light Infantry or Riflemen*. This work was directly based on a new generation of French manuals which incorporated the results of innovative thinking since 1831, especially the doctrines of the *chasseurs à pied*, who from 1838 onwards had been preaching a revolutionary new style of combat. The essence of their ideas consisted of aimed fire at long range with the rifle musket – a weapon which the *chasseurs* themselves had done more than anyone to perfect. Against this, however, they also foresaw a greater role for the bayonet, if only the infantry could be moved more quickly across the battlefield using the jog, or *pas gymnastique*, and if only each man were fully trained in modern bayonet fencing drills. Captain George B. McClellan had in fact already translated these bayonet drills for US use in 1852, but Hardee now put them into context by adding the extra provisions for rifle fire and manoeuvres at a gymnastic pace ('double quick') of from 165 to 180 steps per minute.[32]

Hardee's package was attractively complete and apparently 'modern' in its philosophy. It seemed to chime in well with the thoughts of those American theorists who had picked up the drift of French thinking and who were already predicting a 'rifle revolution' in future battles; yet at the same time it actually retained a very great deal of the original Scott. Admittedly, the line of three ranks was now finally abandoned in favour of two ranks, and the two-man skirmish team was expanded to a four-man buddy group of 'comrades in battle'. Even greater emphasis was placed upon target training and firing from cover, while a jogging unit was expected to cover five miles in an hour, keeping 'as united as possible, without however exacting much regularity, which is impracticable'.[33] Nevertheless, the essential Scott remains as the basis for all the close-order and many of the open-order manoeuvres, including all the complex paraphernalia of company volleys, ployments, deployments and squares against cavalry. Hardee's manual did not therefore turn out to be a revolutionary document after all, for although it recognised that skirmishing was gaining an increasing importance on the modern battlefield it by no means abolished the traditional concept of fighting in line, shoulder to shoulder, two deep.

In the Civil War it was Hardee's manual that won the widest usage and was issued in the most variants. Altogether perhaps a dozen different versions appeared, including extracts deemed relevant to black regiments, a précis for militia use, Union editions which coyly withheld the name of the (secessionist) author, and so on.[34] Of particular interest, perhaps, is the 'Zouave' adaptation by the colourful 'Colonel' Elmer Ellsworth, the 22-year-old founder of the Chicago Zouave cadets in 1859, first commander of the New York Fire Zouaves in 1861, and the first Federal officer to be killed in action – although admittedly the action in question was against only one badly outnumbered Rebel hotelier.[35]

The idea of the Zouave cadets was to take the new wave of French drills – including their full gymnastic rigours – to a peak of perfection which was generally beyond the reach of troops drilled only according to the basic Hardee or Scott. Had he lived, Ellsworth would doubtless have pointed out that in most actual combat the key elements of the advanced *chasseur* theory were being quietly overlooked or dodged, leading to a very different type of battle from that which had originally been envisaged. Thus there was little training for or use of long-range aimed fire by the mass of the Civil War infantry, and double-quick time itself was very rarely used in the manner intended. The rank and file of the American infantry simply could not be persuaded to jog around the battlefield at 165 steps per minute for anything more than short periods, after which they would be too blown to fight at their best.[36] Short of widespread and intensive 'Zouave' training, in fact, those parts of Hardee which might possibly have created a genuine tactical revolution were doomed to remain dead letters.

Actually there was a confusion at the root of all this, since it had really been the French *chasseurs* (metropolitan troops of a scientific turn of mind) who had been responsible for the new thinking, rather than the (colourful, tough, but distinctly unscientific and Algerian) zouaves. Even in France itself such subtleties were not always understood, and we find that in the Italian and Crimean campaigns of the 1850s both these élite units were often squandered indiscriminately in mass assaults, regardless of their true skills.

It is also probably true that Ellsworth and his followers in America were more concerned with the drills, the athletics, the gaudy uniforms and the general showmanship of the whole thing than the battlefield tactics. There was more than a grain of truth in the cynical comments which equated the Chicago Zouave drill with a 'circus manoeuvre',[37] and this interpretation was scarcely weakened either by the circumstances of Ellsworth's demise or by the less than glorious showing of his New York Fire Zouaves at First Manassas. Of still greater importance, perhaps, was the peculiar organisational structure of many of the early Zouave units. The Chicago company had numbered but 61 soldiers plus 15 bandsmen, while the Albany company had a strength of only 66. When the war started these companies retained their identity as independent municipal organisations outside the official army, thereby distancing themselves from the chief sources of recruitment and institutional growth. When their members went to war it was usually either as a small group in a non-Zouave regiment, or as individuals. The Albany Zouaves took pride that no less than 74 of their alumni entered the Union forces as commissioned officers, and this was doubtless beneficial for the general standard of drill and military knowledge in the army. However, these men joined no less than 42 different units, of which 16 were not even infantry.[38] The impact of their specifically 'Zouave' conditioning was therefore dissipated and lost.

In any case, there were peculiar problems associated with any deliberate attempt to build self-consciously élite units in the American armies. A few sharpshooter units were able to win a specialised role by virtue of their exceptional armament, but by and large the truly élite regiments of the Civil War matured by accident and without formal status as such.[39] There was no conscious decision to make the 'Stonewall Brigade' or the 'Iron Brigade' any different from any other outfit – they just turned out that way by virtue of their fine combat performances. Conversely, the intention to make the Zouave regiments into something special ran aground because there was no organisational basis to maintain such a status.

If Hardee's manual can today be criticised in retrospect for failing to impart the full 'Zouave' concept to the Civil War armies, it was more widely criticised at the time for failing to incorporate Scott's third volume – the manoeuvres of an army corps. Several efforts were made to supply this need, but in the end it was the work of Brigadier Silas Casey that was accepted for general use in both North and South. Casey was well fitted for the task, since he had been a member of the board which had originally accepted Hardee's manual, and in the Civil War he had seen hectic action at Seven Pines, then ran the Army of the Potomac's main drill school at Arlington. Many was the green Northern regiment which received its first shock of acclimatisation to military life at his hands.[40] Casey's concern was, however, less with forward-looking battlefield concepts than with the simple practicalities of making existing procedures coherent and easy to digest. The success of his drill book lay in its straightforward style and common sense. Whereas previously there had been some doubts or unclear passages in Scott, now there was system and order. Budding brigade commanders could easily work out where they should place themselves as their regiments marched to the front in column or line, how the brigade artillery should be deployed, how the head of a column should be manoeuvred to give direction to the main mass, and so on. Those were the things they wanted to know, and Casey gave it to them.

Casey also produced a lightly revised version of Hardee's first two volumes, which simplified them at a number of points but changed nothing of fundamental importance. These, too, were generally welcomed for their clarity, soundness and lack of worrying innovations. They took their place beside Scott and Hardee as the books from which the war would be fought.

It was only in 1867, two years after the end of hostilities, that a really 'different' drill book appeared. This was Major General Upton's *A New System of Infantry Tactics, Double and Single Rank, adapted to American Topography and Improved Firearms*. Upton had been one of the more successful tacticians of the war, most notably at Spotsylvania, and his work represented the first formal admission by the US Army establishment that the manuals used during the war had left something to be desired. It is interesting too, because it shows us how he thought the war *had* been fought. His very title expresses an impatience with European models, when it comes to fighting in the difficult terrain of the young continent, and it is notable that his is the first official manual that was not directly based upon a foreign prototype. Upton's title also reminds us that in 1865 the US Ordnance Board had formally adopted the breechloading rifle as the standard infantry weapon, thereby casting a veiled reproach upon the muzzle-loaders with which the war had been fought. One aim of the new manual was therefore to modernise tactics in conformity with the changed character of battlefield firepower.

As with Hardee's and Casey's, however, we find that Upton's manual is ultimately somewhat less innovative than its propagandists might lead us to believe. Once again the title reveals the content, because we find that this is a system of drill based upon 'double and single rank' – or in other words very much the same formations as Hardee himself had used. The difference is only one of emphasis. Upton moves nearer than Hardee to the idea that battles will always be fought in single rank or skirmish order, and he lays more emphasis on heavy volumes of fire delivered by four-man groups. He also talks of developing the morale, presence of mind and individuality of each man so that he can adapt his own tactics to the circumstances – a thought which was very much in keeping with the general trend of military theory in the nineteenth century.[41] Yet Upton retains Hardee's double-quick march in formation, he retains manoeuvres directed by a colour held aloft at the centre of the regiment, and (more in keeping with Scott than Hardee) he comes out against fire at long range in favour of fire withheld until 'deadly range'.[42] The square against cavalry is also retained, as is the idea of fighting in formed ranks:

The officers will observe that a too scupulous regard for cover will make the men timid; they should therefore set the example of fearless exposure whenever an advantage can be gained.[43]

There is still a Napoleonic smack about all this which really debars Upton from the title of 'the father of modern tactics'. His four-man teams do not infiltrate through an enemy position like the German storm troops of 1917, but continue to rally round the flag like men of days gone by.

Perhaps the most 'modern' aspect of Upton's manual is the fact that he talks about *tactics* – the use to which drills can be put in various combat situations – as well as about the drills themselves. Previous drill books had assumed that the key tactical decisions would come easily to a commander,[44] and that all he would need to know would be the details of how to issue orders phrased so that his subordinates would instantly recognise what was required and then all move together to execute it. Whereas these executive orders and movements were standardised and codified, the uses to which they could be put were not. With Upton, by contrast, we find some quite detailed recommendations on how an attack should be orchestrated, with formations and movements listed for a pinning force in front, a manoeuvre element rolling up the flank and a succession of reserves in the rear.[45] This makes his book a true tactical manual for the brigade or division commander such as not even Casey's had been, and we can see that it represents an attempt to pass on a rather different type of practical experience from that described in earlier books.

Upton's subtle shift of content from drill to tactics is certainly indicative of dissatisfaction with existing manuals, in one who had experienced and mastered the searing blast of warfare at ground level. His instinct is to analyse the things which he knows are important but which the books do not discuss. On the other hand it is also possible that this approach was forced upon him by the essential formlessness of Civil War battles. If rigid drill broke down in the tangle of the Wilderness or under enemy fire, then the writers of drill books obviously had to think of something new to put into their pages. With this in mind, therefore, we will turn now to consider the extent to which the formal prescriptions of the theorists were actually put into practice on the Civil War battlefield.

THE VALUE OF DRILL

There can be no doubt that the average Civil War regiment threw itself furiously into the task of learning its drills, and was able to pursue this aim relentlessly for weeks, months and even years before it first came into battle. With a drill parade each day in the morning and afternoon, followed by close study of theory in the evening, the erstwhile civilian was soon brought to a high standard of knowledge in his new calling as a soldier. 'It is drill, drill, battalion drill, and dress parade',[46] wrote one of them, describing his early months in the ranks. There was a clear determination at every level of command that, although the regiments might be improvised in many of their aspects, they would at least become professional in this one. There was a feeling that a crash course in drill could go a long way towards making good the general ignorance and inexperience of military matters from which almost every unit was forced to start.

In certain respects the Civil War armies even enjoyed advantages denied to some regular armies, since the wide social cross-section in the ranks ensured high average standards of intelligence and an enthusiasm to learn. There was also a very salutary emphasis upon drill with large formations – often entire brigades and sometimes even divisions and corps. In the scattered battalion or company-sized garrisons of Britain or France such manoeuvres were a very occasional luxury indeed – just as they had been in the pre-war United States Army.[47] With tens of thousands of men concentrated under canvas on the Rappahannock or the Tennessee, however, there could be massive parades or field days at least once a week. Almost any excuse was good, ranging from a tattoo in honour of the ladies of Richmond to the solemn parade of an army corps to witness the execution of a deserter, from a divisional revivalist rally to the visit of a political leader. Mock battles, snowball fights, inter-brigade drill competitions or simply unadorned 'training' – all of these activities enabled the Civil War regiments to become accustomed to marching in step, manoeuvring together on the word of command and forming a front in unison.

Admittedly, there were many discrepancies in the amount of such drilling any individual unit would receive, and especially in the first year of the war there were some regiments whose colonels failed to attach any importance to it at all.[48] Throughout the war it was very rare for blank ammunition to be used in mock battles, and common enough for any drill to be regarded as boring or demeaning. It was not always of positive value, as witness the case of 19th Iowa in the spring of 1863, when the general feeling was that the incessant drills were simply for the amusement of officers during the intervals between their drinking-bouts.[49] In many ways drill was a highly un-American and undemocratic activity.

None the less, when it came to fighting battles, a sound background of drill training was undoubtedly of great value to a regiment. In particular it imparted two specific benefits which may be described, respectively, as 'tactical articulation in the period leading up to close combat', and *'esprit de corps* once the serious killing had begun'.

By 'tactical articulation' we mean the ability to move large numbers of men quickly and without fuss from one place to another, in conformity with the terrain and the demands of the battle. Clearly this can be done far more effectively if everyone follows a standardised procedure than if they simply mill about as a crowd. If they follow a drill, furthermore, their commander has at least a chance of predicting the way they will move and the time they will take. Even if they seem to take rather a long time, he will still be able to command more effectively if he can predict it in advance. Thus '60 minutes to form a Division into line of battle' may sound like an interminable age to perform a very simple movement, and '30 minutes to form up in a field after moving through a wood'[50] may sound even less snappy – yet it is calculations like these that lie at the very heart of generalship. Without a standardised drill, furthermore, the time taken in each case would probably run into hours rather than minutes.

In extreme situations the use of drill techniques for the ployment or deployment of a force could mean the difference between defeat and victory. Thus at Peach Tree Creek 123rd New York was almost caught out by the Rebel attack while its brigade was closed up to take lunch. Only by deft manoeuvring was a firing line established, in the nick of time:

If every regiment had not been composed of experienced veteran soldiers, men who knew what to do under the most adverse and changing conditions, we would never have fronted in battle line in time to have made a successful defense ... [51]

Something similar happened to Johnson's Division of the Confederate Army on Mine Run in 1863. The division was marching along a road through a wood when it suddenly suffered a surprise flank attack by Union skirmishers. Undeterred, it rose to the occasion with scarcely a pause for thought:

Regimental officers cut off companies from their regiments, formed them as skirmishers right in the road, and ordered them forward. I must say this was the promptest movement I saw during the war.[52]

A further example of the value of drill in the preliminaries to combat comes from Antietam, when 35th Massachusetts was ordered to advance against an enemy which was not directly to its front:

... Colonel Carruth started up: 'Attention! Thirty-Fifth.' We rose up at once and faced the front, forming forwards a little, the companies moving to their positions. 'Left-face! Forward-march!' Hardly had the regiment faced and moved a little distance when the [retiring Union] battery came dashing full speed into us, breaking our line for a moment, but the men undismayed closed up immediately. A little way to the left, then facing to the front, with a hurrah, the regiment went at double quick, in line of battle, over the hill and down the slope into the valley towards Sharpsburg.[53]

Tactical articulation of this sort could sometimes continue for some time even after a regiment had come close to the enemy and the casualties had started to mount. With a reasonably well drilled force it was often possible for an initial volley to be delivered 'according to the book', or for a change of formation to be effected in a moment of crisis. Time and again we find diarists expressing their awe and admiration at the regularity of some movement performed in the heat of battle as if on a parade ground, although we may also perhaps speculate that at least a part of this impression came from one of two sources of subjectivity or bias.

The first source of subjectivity is a natural readiness to be excessively impressed by the enemy, and even to attribute mythic powers to him. On many of the occasions on which observers reported 'parade-ground' drill in battle it was in the enemy's ranks that they discerned it, rather than in their own. Thus Colonel Polk of the North Carolina militia tells us that in December 1862 as he watched a Union advance against him: 'I could not help admiring the beautiful order and measured tread of the villains as they neared us.'[54] Equally at Fredericksburg in the same month William Owen of the Washington Artillery of New Orleans says that as the Northern troops came into the attack

The jingling and rattling of accoutrements and the commands of the officers, 'Forward! Forward! men! Guide center!' were quite distinct ... What a magnificent sight it was! It seemed like some huge blue serpent about to encompass and crush us in its folds. The lines advanced at the double-quick, and the alignments were beautifully kept.[55]

One of the lesser functions of drill was certainly to reinforce the impression in an enemy's mind that your men had an irresistibly high standard of organisation and training.

The other source of bias which may have unduly affected the impression of drill-ground precision in battle lay in the admiration felt for certain élite units, both at the time of their battles and retrospectively, when the moment came to write about them for later generations. For example, at Chancellorsville Rice C. Bull says he witnessed a 'wonderful' movement by 2nd Massachusetts, firing as it advanced across a cleared field and then as it retired, marching backwards in perfect order. Yet he concedes that 'This regiment was considered one of the best drilled in the Army'.[56] Again, from Murfreesboro we have an account of another crack formation – the celebrated Orphan Brigade – which shows how even some of the small

107

print of Hardee's second volume could be observed during an attack under crippling fire:

... with steadiness of step we moved on. Two companies of the Fourth regiment, my own and adjoining company, encountered a pond, and with a dextrous movement known to the skilled officer and soldier was cleared in a manner that was perfectly charming, obliquing to the right and left into line as soon as passed.[57]

Another example comes from 13th Massachusetts at Antietam, which advanced in double column at half distance under artillery fire, zigzagging by a series of direction changes in order to reduce the casualties:

This movement made under a heavy fire was performed with as much precision and coolness as though the regiment was on a battalion drill. It is worth mentioning to show what good use may be made of the skill and confidence acquired by constant drilling.[58]

When this regiment finally became engaged in a close range firefight, however, it simply stood its ground and fired off all its ammunition without a further manoeuvre. At that point its soldiers ceased to use their skill in drill movements, even though the confidence it gave them continued to buoy them up and keep them in the battle. This change marked a subtle but important transition from the tactical articulation of the approach march to the inarticulate action of *esprit de corps* in the combat itself.

The same transition can be traced in the experience of 24th Michigan at Fredericksburg. At first the regiment was manoeuvred *en masse* in the van of Doubleday's brigade column, then on taking position on the left flank of the army the brigade deployed into two lines, including a deployment of 24th Michigan into a single regimental line. This line advanced 'as straight as on parade, walking around obstacles and immediately reforming'. It crossed a field and a wood, its skirmishers driving the battle before them. A regular wheel brought the regiment forward onto Massaponax creek to face the enemy as the fire became heavier.

So far everything had gone smoothly according to the drill book's function as a lubricant of prompt tactical movements. At this point, however, the coherence of 24th Michigan's action started to slip away. Some men went to ground under the heavy fire. Others wavered as the colonel of the regiment waited for orders from the brigadier. There were no positive manoeuvres to be performed, but to keep the men occupied the colonel started to shout orders from the drill manual: '"Attention battalion! Right dress! Front!" There on the open field he put his men through the manual of arms!'[59] By this apparently eccentric behaviour the colonel was actually reminding his men of their training and their identity as a disciplined body. This was the *esprit de corps* function of drill taking over where the movements left off. It acted at the level of the subconscious rather than of the strictly rational, but it could still be a powerful influence in calming and heartening a unit under stress.

Captain De Forest reports that in his first battle the main value of drill lay precisely in this function as a sort of talisman or good luck charm:

In our inexperience we believed that everything was lost if the men did not march shoulder to shoulder, and all through the battle we laboured to keep a straight line with a single-mindedness which greatly supported our courage.[60]

Yet another example, albeit a rather more light-hearted one, came in 5th New York at Beaver Dam Creek, as a fearsome Confederate assault was approaching:

Our colonel indulged in a grim bit of humor. 'Attention, battalion!' he shouted. 'Parade rest!'

The order was promptly obeyed, though the men laughed to see the regiment thus put through holiday manoeuvres in sight of the enemy. Our colonel's coolness, however, had its intended effect, for other moving columns stiffened up and passed on in excellent shape to the position assigned them.[61]

There were of course many other influences which could help a unit to keep its steadiness in battle. Any meaningless or inane murmurings by an officer who seemed self-assured could work to great effect, as Captain Oscar Jackson of 63rd Ohio richly understood in his battles around Corinth:

It is customary in engagements to have some motto or battle cry given by some commander, such as 'Fire low', 'Stick to your company', 'Remember some battle', (naming it). To draw the attention of my company while the [enemy] charge was advancing I said: 'Boys, I guess we are going to have a fight.' This is always a doubtful question to an old soldier until he sees it, but they all believed it this time. 'I have two things I want you to remember today. One is, we own all the ground behind us. The enemy may go over us but all the rebels yonder can't drive Company H back. The other is, if the butternuts come close enough, remember you have good bayonets on your rifles and use them.' And well did they remember what I said.[62]

Had he authoritatively shouted 'Boys, the future lies ahead', Captain Jackson would doubtless have drawn an equally heroic response from his men – and many indeed are the military, political and religious leaders down the ages who have taken this truth, alas, to heart. The important thing about leadership is often not its content but its style.

The buddy group is also of enormous importance to unit steadiness in battle, since the soldier fights for his company, his squad and his 'comrades in battle' far more than he fights for such abstracts as liberty or defence of the homeland. He sticks with the colours because he says to himself, 'Here is Bill; I will go or stay where he does.'[63] If Bill seems for the moment to be staying with the regiment, then our soldier will stay too, and by extension the whole regiment will stand its ground. Conversely, if Bill runs away our soldier may feel that he too is free to run, and so by example the panic will spread very rapidly.

It must remain an open question whether or not all this was fully understood by the authors of the French *chasseur* drills upon which Hardee based his 'comrades in battle', but it would certainly appear from modern research into combat stress that between them they had hit on a winner. The way they liked to express it at the time was by the French word *surveillance*, meaning the way in which older soldiers would watch over the younger ones, and the younger ones would watch each other. Obviously *surveillance* was far easier in closed ranks than in open order, and it always provided one of the most powerful arguments against the increased use of skirmishing. Not to put too fine a point on it, you could ensure that men stood and fought – and died – if you had them all enclosed in serried ranks, whereas you could have no such certainty when they wandered off as individual skirmishers to lie down behind their own pieces of cover, away from the regiment. This was how Captain Nisbet of 21st Georgia dealt with a somewhat shaky soldier who eventually deserted to join the Union army:

Before he deserted, however, I got him into three fights by putting him in a file with good men and ordering them to watch him and see that he did not drop out ... I would not let him stop on any excuse. When the firing commenced he started back. I grabbed him by the collar and ordered him forward, threatening to shoot him. He said: 'Captain, my gun's stopped up.' [64]

Surveillance was always intended as a 'family' matter, within the informal community of the company or the squad. Harsher and more serious could be the battle police imposed by higher authority to round up fugitives from whole brigades or divisions. This was Grant's recommended use for his cavalry at Shiloh, and Lee's use for 21st Georgia after it had been relieved from the firing line at Antietam. At Corinth the hated martinet, Bragg, detailed one-tenth of his troops to patrol the rear of his army and shoot deserters, while at the start of Gettysburg the depth of Meade's anxiety may be judged by his general order authorising the 'instant death of any soldier who fails to do his duty at this hour'.[65] Not all the executions of deserters, apparently, took place in cold blood or after due process of law.

Drill must be seen in context, as one among many influences by which a unit's officers might hope to maintain steadiness and *esprit de corps*. It took its place alongside such carrots as long familiarity with the unit, the personal example of buddies, the calming words of officers – or simply a good breakfast[66] – and such sticks as the formal or informal retributive measures which might be taken against the battle-shy. Drill training provided only a *background* of discipline and obedience, which could

110

sometimes help to keep troops attentive to their duty. What it could not usually hope to provide, however, was precision of movement once close combat had been joined.

THE BREAKDOWN OF DRILL

General D.H.Hill in later life once asked, rhetorically but memorably: 'Whoever saw a Confederate line advancing that was not crooked as a ram's horn? Each ragged rebel yelling on his own hook and aligning on himself.'[67] From the other side of the Mason-Dixon Line General Sherman also pointed out that

Very few of the battles in which I have participated were fought as described in European text-books, viz., in great masses, in perfect order, manoeuvering by corps, divisions, and brigades. We were generally in a wooded country, and, though our lines were deployed according to tactics, the men generally fought in strong skirmish-lines, taking advantage of the shape of the ground, and of every cover.[68]

Such testimony as this is probably enough in itself to prove that whatever else drill may have done for the Civil War soldier it scarcely guided his actions all the way through a battle. At some point drill would break down and the individual would be thrown back on his own devices. In all probability this had also happened in the classic Napoleonic battles 'as described in European text-books'; but, even if it had not done so then, it certainly did so in the 1860s.

In part this disintegration was a question of the terrain, as Sherman implies. Often also it came from a conscious decision to employ skirmish order – that paradoxical drill formation which was itself the deliberate abandonment of drill. We have already seen that each successive American infantry manual from 1812 to 1867 placed a greater emphasis than its predecessor on skirmishing. In essence, however, the most important cause of the disintegration of drill lay in the proximity of the enemy, and therefore in the spread of excitement, fear and unusual happenings of every kind.

We can scarcely improve upon the following oft-quoted descriptions of Antietam from 9th New York:

It is astonishing how soon, and by what slight causes, regularity of formation and movement are lost in actual battle. Disintegration begins with the first shot. To the book-soldier all order seems destroyed, months of drill apparently going for nothing in a few minutes.

The truth is, when bullets are whacking against tree-trunks and solid shot are cracking skulls like egg-shells, the consuming passion in the breast of the average man is to get out of the way. Between the physical fear of going forward and the moral fear of turning back, there is a predicament of exceptional awkwardness from which a hidden hole in the ground would be a wonderfully welcome outlet.[69]

All else being equal, the loss of regularity in a regiment will occur at the moment when this 'predicament of exceptional awkwardness' begins to be experienced.

Sometimes the disintegration occurs simply by accident, as one man's innocent word or gesture is misunderstood and turned into another man's panic. Sometimes it is a matter of badly delivered orders, or the sudden fall of an officer responsible for giving orders. The disintegration may happen gradually, as a firing line two deep imperceptibly converts itself under pressure into a strong skirmish line, or as a 'column by division at half distance' slowly but ineluctably turns into a 'howlin' mob'. Sometimes the disintegration is resisted nobly for a while by some particularly well trained or unduly sober unit, and sometimes it is embraced with almost indecent relief by a poorly trained or inebriated unit (the *enemy's* use of whiskey to supercharge his attacks was noted on many occasions by shocked observers from both North and South). Sometimes drill evaporates reluctantly and under extreme provocation; sometimes it vanishes instantly of its own accord, as in the following passage from 55th Illinois under attack at Shiloh:

Soon we were called to attention, line formed and wheel by company ordered. This seemed too much for the men's overwrought nerves, and as the wheel was started the entire line broke and ran.[70]

On this occasion the breakdown of drill had the effect of creating a panic rout, but in many other cases it simply led to an abandonment of the precision and 'etiquette' assumed in the manual, even though each soldier continued to fight and to do his duty as he saw it.

When a line opened fire with a command volley it was quite normal for the men to continue firing at will, independently, after the first shot. Indeed, this type of fire had been recognised as the most practical in battle by every drill book since at least 1791, providing yet another case in which the formal drills themselves demanded the abandonment of regularity.[71] Unfortunately, however, it also encouraged disintegration, because once fire at will had started it often proved very difficult to stop. Scott had recognised this problem in his manual, as had every other experienced officer who stopped to think about tactics, but the Civil War record seems to suggest that no effective solution was ever found. Time and again we read of regiments blazing away uncontrollably, once started, and continuing until all ammunition was gone or all enthusiasm spent. Firing was such a positive act, and gave the men such a physical release for their emotions,[72] that instincts easily took over from training and from the exhortations of officers – whose voices would in any case be drowned out by the din. Thus it was often the moment of opening fire that represented the moment when command, control and parade-ground drill all disappeared together.

By the same token it was often very hard to stop individuals – or even whole units – from opening fire before they were ordered to, once the stresses of combat had started to be felt. Alternatively, these stresses might lead a regiment to charge recklessly forward, against orders, simply to 'do something positive' against the enemy.[73] Yet again, the men might lie down and take cover spontaneously when under fire, perhaps creating an unplanned ambush for an enemy who could not identify their position. In all these and many other ways the breakdown of drill could take place without a unit actually having to leave the battle line or abandon its efforts against the foe.

The breakdown of drill might also take the form of a breakdown of the prescribed organisational structure, and it was not unusual to find regiments, companies and individual soldiers fighting outside their normal order of battle. Thus at Resaca in 1864 James Nisbet formed a line against the Union attack with his own regiment and two more which just happened to be passing. In effect he fought with an *ad hoc* or improvised brigade until he was relieved by a formally constituted brigade of Texans.[74] Then again in the Wilderness battle at about the same time we read of the utter confusion which the heavy woodland created in the Union line: 'From Hancock down through Birney and Gibbon, each general commanded something not strictly in his command.'[75]

Then again, at Corinth in 1862: 'The 18th Missouri was not even pretending to function as a regiment that day; its men were fed into the line wherever McKean [the Division commander] needed firepower.'[76]

In the three cases cited the mix-ups were deliberately accepted by higher commanders, but at many other times there was a quite unwanted confusion of units. This happened most notably, as we have already seen, in the passage of lines. At Antietam 6th Wisconsin became engaged with the enemy and was then joined by its second line:

The Fourteenth Brooklyn Regiment, red-legged Zouaves, came into our line, closing the awful gaps. Now is the pinch. Men and officers of New York and Wisconsin are fused into a common mass, in the frantic struggle to shoot fast. Everybody tears cartridges, loads, passes guns, or shoots. Men are falling in their places or running back into the corn ...[77]

Such complete confusion was the rule rather than the exception in Civil War battles,[78] as doubtless it is in the battles of every other era of military history. Although the regiments and brigades may start off a battle with their drills, organisations and identities intact, these parade-ground conventions soon become blurred and compromised when the lead begins to fly.

In fairness to the drill books of the Civil War, we must remember that they were not true tactical manuals like Upton's of 1867. Scott, Hardee and Casey were intended only as recipe books for basic movements: they did not pretend to tell their readers much about how to fight battles or how to behave when the enemy was close. They were supposed to provide an auxiliary to – but not a substitute for – such essential training aids as sustained supervision by an experienced *cadre* or battle inoculation with real ammunition. In the peacetime regular armies of both Europe and the Americas these two supplements to the drill book had always been available to a greater or lesser degree, particularly through the experience of small wars in unpromising theatres like Mexico, the Indian frontier, French Algeria or British India. In the Civil War none of this background was available to the improvised mass armies, so the drill manuals were served up unadorned to green regiments with the implied but unspoken message that 'If you read this slim volume you will find out everything you need to know'. Nothing could have been more misleading, and it is scarcely surprising if a considerable wave of ill-feeling against the books was the eventual result.

Much of the criticism was directed against the supposed obsolescence of the manuals, which in turn was attributed to their European and hence, in American eyes, obscurantist or reactionary origin. European terrain was assumed to be significantly different from American terrain, as were the weapons in use – Enfield, Whitworth, and Minié notwithstanding. The spurious technicalities and fancy language of the Europeans and their West Point imitators could also be relied upon to raise a laugh, as in the following story told by Sam Watkins about his (non-West-Point) Lieutenant, J. Lee Bullock:

Lee, anxious to capture a battery, gave the new and peculiar command of, 'Soldiers, you are ordered to go forward and capture a battery; just piroute up that hill; piroute, march. Forward, men; piroute carefully.' The boys 'pirouted' as best they could. It may have been a new command, and not laid down in Hardee's or Scott's tactics; but Lee was speaking plain English, and we understood his meaning perfectly, and even at this late day I have no doubt that every soldier who heard the command thought it a legal and technical term used by military graduates to go forward and capture a battery.[79]

The lack of realism of the drill books was even equated by some, such as De Forest, more or less with fantasy or fiction: 'It is only in pictures and civilian novels that brigades and regiments charge at the double *with even front* over any distance above sixty or eighty yards.'[80]

Some officers stated that the rigidity of many of the drills often stood in the way of tactical articulation, particularly against flank attacks, even though this was precisely the thing the drills were supposed to do best. Others felt, as many Napoleonic officers had felt in the past, that all but a few basic movements were irrelevant to the real battlefield, so the training curriculum should be significantly reduced and simplified.[81]

Most of these criticisms could probably have been met if some effort had been made to supplement the drills with a true tactical manual and to give a full explanation of just what the drills themselves could, and could not, be expected to deliver. With hindsight we can say that both Richmond and Washington ought to have appointed a permanent committee to review and evaluate tactical doctrine, constantly updating and rewriting the manuals to suit the changing needs of the armies. In the late nineteenth century this concept was starting to appear in the Prussian Army, but it was still somewhat too futuristic and imaginative for soldiers elsewhere.

It is at least not really fair to blame the drill books for describing an obsolete European type of war, since even the innovative American Upton was still thinking in very much the same terms as the Europeans two years after Appomattox. The European drills, especially those based upon the *chasseurs à pied* theory, really reflected the most modern tactical ideas available anywhere in the world at that time. The reason why they seemed to be unsatisfactory was rather that they were placed, without further clarification, into the hands of an exceptionally well educated army which had exceptionally few experienced officers. The result was that too much was expected from them, and too little of their wider context was understood or explained.

Perhaps, as Watkins hints, it was really all a matter of language and style. In modern times armies employ Madison Avenue copy-writers to present their manuals to the common man in the idiom of the comic strip or the commercial, so maybe something similar ought to have been done by Hardee before he sent his translated French text to the printer. On the other hand, it may be that the problem really had more to do with the eternal dilemma facing all writers of military manuals. They know they must offer a recipe for action, yet they also know that in real battle this recipe will quickly fall apart. There is probably never any realistic possibility of sustaining drill movements – whatever their precise content may be – beyond the approach march and the start of close combat. Once that point has been reached there is no power on earth which can stop 'each ragged rebel yelling on his own hook and aligning on himself'.

5

The Battlefield
and Its
Fortifications

The men took positions behind a curving line of rifle pits that had been turned up, like a large furrow, along the line of woods. Before them was a level stretch, peopled with short, deformed stumps.

The Red Badge of Courage, p. 75

The breakdown of regular drill may be attributed primarily to the imminence of danger and the extreme emotions provoked by it. Many commentators have pointed out, however, that the particularly difficult and broken terrain of North America was also a powerful factor in the process. Just as the deep forests in the Revolutionary War supposedly gave a great stimulus to light infantry skirmishing, so the wildernesses of the Civil War may have produced a similar result. The soldier may have welcomed the chance to fight as an individual because of the rugged independence of his race – but he was also forced into it by the trees.

Against this view, on the other hand, we have the diametrically opposite image of formed lines advancing over open meadows, suffering casualties from accurate rifle shots at long range. If the men break ranks and spread out as skirmishers, according to this view, it is in response to new weapons with wide fields of fire. As skirmishers and marksmen, furthermore, they will themselves be better able to use their own weapons to full effect at long range.

We can now look a little further into the physical aspects of the battle-field in order to discover which tactics may have been possible and which were not. So many differing claims have been made in this connection that a little clarification is surely timely.

THE TERRAIN AND THE WEATHER

Of course there is no such thing as the 'average' Civil War battlefield, since some were heavily wooded while others were open and bare of cover. Some of the battles were fought in the swampy bayous of the Mississippi, others around the parched water holes of Kentucky. A few took place in built-up areas, on steep rocky crags above the clouds, or even in tunnels beneath the ground. The spectrum of different terrain types was almost as wide as that seen in the Napoleonic Wars themselves.

In the Eastern theatre, however, we can say that the major battles were usually fought on relatively open farmland interspersed with small woods and the occasional stone or timber-framed building. Even where a battlefield included a tangle of difficult undergrowth, such as the sodden malarial swamps of the Chickahominy or the rocky pine-clad Round Tops at Gettysburg, there were normally open fields of fire close at hand. Generals adopting a defensive posture could usually find some sort of ridge line with a commanding view, although this would rarely be free from covered approaches which an attacker might try to exploit.[1] At Second Manassas Jackson anchored his line on an unfinished railroad embankment which ran partly through open fields and partly through a wood, and it was the wood which saw the heaviest fighting. At Fredericksburg his position was remarkably similar, although he appears to have used it more in depth – perhaps deliberately turning the wooded parts of his line to the advantage of his own counter-attacks.[2] In the same battle the Confederate left wing dominated a completely open glacis almost half a mile wide, in which the only cover was a fold in the ground quite close to the defenders. Once the attackers had reached this feature they found that they could not (or would not) make further progress to the front because of the strength of the Confederate position, yet were equally exposed if they tried to retire to their start line. They had no option but to go to ground and stay there.

At Malvern Hill the Confederate attackers were able to form up under cover of a wood, but they then faced about a thousand yards of open fields covered by massed Union batteries on rising ground behind. At Antietam the battlefield was more undulating, with a scattering of copses and hollows providing some cover to the attack, although the defenders could still find plenty of good positions with commanding fields of fire.

Thus far we can agree that many of the important Eastern battlefields lent themselves admirably to formal close-order drill, but also to long-range fire from both artillery and infantry. The fact that the closest fighting tended to be channelled into the closest country might also be cited as an indicator that long-range fire was effective: if troops cannot move across the open beaten zones, they will seek the protection of woods.

An alternative explanation is also possible, however, since in open country the ebb and flow of battle can be faster. Commanders can see their chance of success or failure more easily, and therefore do not have to put it to the test of close combat. Regardless of weapon ranges, a beaten defender will be able to run sooner, just as an outfaced attacker will realise earlier his mistake in advancing. In conditions of low visibility in woods, however, the result cannot be known until every man has put it to the test of almost individual combat – a process which can be long and costly. Hence, if most casualties in Civil War battles occurred in thickets, that may be the result more of low visibility ranges within them than of high weapon ranges outside.

We should also remember that even the most 'open' Civil War battlefields actually contained a great deal of cover for the infantryman who went to ground. Just as Wellington had concealed his army at Waterloo in an apparently clear field, so the American armies habitually covered themselves quite adequately behind such features as a light rail fence, an almost indistinguishable dip in the ground, or even the flimsy protection of standing corn. Each of these could provide a starting-point for a defensive line occupied by soldiers lying prone. If there was a sunken lane, a stone wall, or some similarly solid bulwark, then that would be better still. For an army which could afford to remain static and was not too proud to lie down, almost any piece of ground could quickly be converted into a stronghold, even at close range from the enemy. One did not need a mountain, a forest, or a river line in order to gain security.

In some parts of Virginia there were certainly grandiose natural features at the disposal of the Civil War armies. The great rivers channelled strategic movements and were occasionally the scenes of assault crossings under fire. In the Blue Ridge mountains there were some dramatic slopes to be stormed by any force wishing to control the highways which ran through the gaps. In terms of the large-scale battles, however, the most important natural feature of all was the great dank forest which sprawled along the southern bank of the Rapidan. Its reputation had already been ominous before the Battle of Chancellorsville in the spring of 1863, but by the time the battered armies emerged from it a second time in the summer of 1864 the very name of 'this repulsive district'[3] seemed to drip with blood. That name was singularly appropriate, and can evoke much of the horror of the war as a whole. It was, of course, the Wilderness.

The Wilderness was the chief exception to the general rule that the battles in the East were fought on partially open terrain. Admittedly, it contained quite a number of firebreaks and cleared smallholdings, just as the more open battlefields contained their share of copses and woods. But the essence of the Wilderness was the impenetrability of its thickets. Firefights could take place at a range of twenty yards[4] without the two sides being able to see each other for the undergrowth, and we read that 'Only the roar of the musketry disclosed the position of the combatants to those who were at any distance'.[5] The density of these tangled woods enabled entire army corps to remain hidden from their foes even when very close; it prevented the massed use of artillery or the coherent exercise of command; it led to the accidental shooting by their own men of two of the Confederacy's ablest generals – Jackson in 1863 and Longstreet in 1864. As an acme of unpleasantness, furthermore, it produced a series of forest fires which engulfed the wounded and anyone else who was unable to get out of the way.

No terrain in the Napoleonic Wars had combined quite the same elements of horror as the Wilderness, although the layout of most of the other battlefields in Virginia, Pennsylvania and Maryland would surely have been easily understood by a Napoleonic soldier. Veterans of the 1814 French campaign would certainly also have been at home with one of this area's most obvious features for the marching man – the 'gumbo',[6] or clinging, treacherous mud. As General Chamberlain explained of the area south west of Petersburg, 'The soil is a mixture of clay and sand, quite apt in wet weather to take the character of sticky mire or of quicksands. The principal roads for heavy travel have to be corduroyed or overlaid with plank.'[7] After heavy rains a bottomless road could halt military movements almost entirely. Burnside's notorious 'mud march' in January 1863 was the most spectacular example of this, but it seriously hampered many of the other operations during the period of high soil saturation between December and May.[8] In Maryland and Pennsylvania there were particularly good metalled roads,[9] but elsewhere in the theatre there was always a problem with mud during wet weather. In the dryer parts of the summer, by contrast, the soil was likely to turn to fine, choking dust, causing exceptional discomfort to anyone unfortunate enough to be caught up in a marching column.

In winter the battlefields could be chillingly cold and unforgivingly wet underfoot. The soldier who threw himself to the ground to escape the enemy's bullets would be fortunate if he could dry himself out within the day, and the powder in his cartridges would be rendered useless if it once became moist.[10] Although percussion priming did certainly give better results when damp than the flintlock system, it was only the metallic cartridges used in some breechloaders that could be considered truly waterproof. On one famous occasion at Peach Tree Creek these arms were successfully fired by troopers from a position submerged in a river,[11] but that was – to say the least – an unusual event.

Bad weather put a damper on the movements, the bivouacs, the health and the firepower of the Civil War soldier; it was little wonder that he liked to go into winter quarters for the worst two or three months of the year. When he could get away with it he would come out to do battle only around the end of April, and even that was still a period of uncertain weather.

Spring and summer in Virginia could nevertheless be idyllic seasons, although to the sweating soldier on the march they might often seem to be too much of a good thing. The blazing, sultry heat could be debilitating, especially if water was rationed, while the dehydration caused by intense action under combat stress could quickly fray tempers, judgement and accurate drill. If one wishes to identify reasons for the breakdown of regularity on the Civil War battlefield, one should therefore surely examine the weather quite as much as the various other factors which have from time to time been suggested.

Before we leave the Eastern battlefields, we will take a quick look at them as they must have appeared to the artillerist and the cavalryman. To gunners they were often less than ideal, since their complex network of small features tended to break up the full effect of massed fire. On some occasions there was a sufficiently open area for concentrated artillery to be devastating – Malvern Hill, Second Manassas, Gettysburg and Petersburg spring to mind. But in many other cases the targets for a grand battery were so fleeting, so far away or so well protected that the guns could be useful only in low-level close support of individual infantry units. This was a frustrating throwback to an earlier era of artillery theory, for which the terrain in Virginia and at Antietam must bear a certain proportion of the blame. In the Wilderness the situation was (naturally!) far worse. In the 1863 battle it was found that few of the guns could be deployed off the roads and, apart from a couple of notable exceptions to the rule, they were less effective than hoped. In the 1864 battle Grant actually ordered home no less than 122 of his guns,[12] which was a quite unprecedented move for any commander in this – and perhaps in any other – war.

As far as the cavalry is concerned, the Virginia theatre was even more disappointing. Whatever opportunities it may have offered for daring headline-grabbing raids on the enemy's rear, its battlefields certainly possessed none of the endless unencumbered plains known to Nansouty and Murat. Even if well trained massed cavalry formations had existed – which they generally did not – they would have found it relatively difficult to land a massed charge at a telling place or time. This was recognised early in the war by the official Union analysis, which identified the Eastern theatre as unpromising cavalry country by reason of its agricultural enclosures and woods. The Western theatre was assumed to be much less obstructed, and given a much higher priority in the allocation of resources for the mounted arm.[13]

In the West the great distances did, in fact, allow the cavalry to assume a greater strategic role even than it possessed in the East, but paradoxically most of the big Western battles were fought in more difficult, rather than more open, terrain. As far as the tactical role of cavalry is concerned, therefore, we can agree with Benjamin McIntyre of 19th Iowa, writing from the battlefield of Prairie Grove: 'Cavalry fights are as scarce here as on the Potomac and with few exceptions they seem mere vampyres hanging on the infantry – doing but little fighting but first in for the spoils.'[14] Unlike the majority of those in Virginia, most Western battles took place in thick woods. In the early operations along the Mississippi and Tennessee rivers it was a matter of the luxuriant, well irrigated growth which fringed each landing stage or riparian fortress. At Shiloh the Confederates achieved surprise because the woods concealed their movements, but then found that the trees helped to disorganise the assault while offering protection to the hard-pressed defence. At Fort Donelson the woods surrounding the Confederate earthworks provided them with a network of abattis, and helped to add a touch of confusion to the fights which spilled out of the work. At Murfreesboro, equally, there were thick cedar groves to impede the movement of an attacker.

In the later Western battles the nature of the vegetation changed somewhat, although it usually remained just as dense. Between Chattanooga and Atlanta there stretched a hundred miles of pine forest through which the armies marched, fought and countermarched. At Chickamauga and Chattanooga the woods helped to disorient the defender and shelter the attacker, although this was soon to change. Sherman reports that

We were generally the assailants, and in wooded or broken countries the 'defensive' had a positive advantage over us, for they were always ready, had cover, and always knew the ground to their immediate front; whereas we, their assailants, had to grope our way over unknown ground, and generally found a cleared field or prepared entanglements that held us for a time under a close and withering fire.[15]

Also in this campaign were at least three major fights on or near very steep slopes – at Lookout Mountain, Missionary Ridge and Kenesaw Mountain. All three features were such formidable natural obstacles that by rights they ought never to have been taken, and if the first two actually did fall it tells us more about the poor morale in Bragg's army than it does about the problems of terrain. Conversely, Sherman's error in assailing the Kenesaw Line is explicable only in the light of these misleading earlier successes, not in terms of the imposing contours which he faced.

Along the southern edge of the Western theatre lay some uninviting swampland battlefields around New Orleans, Savannah and Goldsboro. In several of these actions the troops waded into action with water up to their knees and their sense of direction blurred by the undergrowth.[16] Along the northern edge of the Western theatre, by contrast, lay some battlefields which displayed a mixture of terrain features more reminiscent of Virginia. Agricultural land, small copses and scattered dwellings were more common in the interior of Tennessee and Kentucky than they were along the great river lines or on the road to Atlanta. From a tactical point of view the battles at Perryville, Knoxville and Nashville thus had somewhat more in common with those of the East than did certain of the other Western fights.

West of the Mississippi there were some small actions fought on various different types of terrain, but these may be considered sufficiently marginal to our present concerns for us to follow the South Carolinan Haskell and – perhaps unfairly – dismiss them in the following terms:

There was so much less fighting in the Trans-Mississippi area, that, like the one-eyed man in the country of the blind, anyone who was there was a great leader, and heroes were almost as cheaply made as in the Cuban War.[17]

Instead of examining these Lilliputian conflicts, we will return to the Tennessee and Georgia area, and consider the potential of the terrain as it must have appeared to an artillerist.

As with the cavalry, the close country of the Western battlefields offered but few opportunities to the gunners. It imposed short fields of fire and relatively low mobility, thus making it unusual for artillery to be deployed *en masse* in a decisive manner. Long-range overhead fire by the Union's rifled cannon was often possible in hilly terrain, but this was more a weapon for sniping – as the unlucky General Polk discovered on 14 June 1864 – than for the Napoleonic 'artillery charge' to close range. At close range the terrain was too encumbered and broken for such a tactic to be used, even if sufficient 'tactical articulation' had existed in the Western artillery for such a fluid conception. Instead, the guns tended to be split down to local levels of command and employed piecemeal.

As we have already noted, the logistic difficulties were far more acute in the West than in the East, and the mud liable to be no less inimical to the movement of wagons. Armies of any size were very closely bound to the river or rail lines for their bulk supplies.[18] Since the density of these lines was less than in Virginia, the location of battle sites was determined more completely by them. The density of settlement and farming was also much less in the West, which in turn meant that these battle sites tended to have a savage and uncultivated aspect. It was in the West, in fact, that 'wilderness' was the rule rather than the exception.

THE THEORY OF FIELD FORTIFICATION

No account of Civil War battlefield terrain would be complete without an acknowledgement that the natural qualities of the ground – whether close country or open fields – were often radically altered by the works of man. The battlefields as they were fought over were not always recognisably the same landscapes as had existed in the early spring of 1861. Nor were the changes just a matter of routine pioneering to enhance communications by building bridges or corduroying roads. It was often a case of constructing large-scale fortifications – an attempt to remodel the whole shape of the ground and its combat value.

Before 1861 the US Army had shown that it was already one of those armies that liked to dig itself in. Like the legionaries of ancient Rome, the Swedes of Gustavus Adolphus or the Russians from Eylau to Sebastopol, it had a certain tradition of scattering fieldworks over its battlefields.[19] Indeed, the nation itself had been built by those who had dug the defences of Bunker Hill, New Orleans and the Alamo. There was certainly no stigma of cowardice or lack of initiative attached to such labour – as there was in some other armies or even as there was to be among some American forces of a later era.

It was perhaps significant that the Republic's only official military academy had been built as a college of engineering – a version of Napoleon's École Polytechnique which was never balanced, like his, by a Saint-Cyr and a Saumur. Graduates of West Point might not have been particularly well versed in the art of commanding infantry or cavalry in battle, but whatever else one might say about them it is certain that they had a solid grounding in mathematics and military engineering. Their Professor of Engineering and the Art of War, Dennis Hart Mahan, was to all accounts a persuasive teacher – and his favourite theme was the pre-eminence of the spade in combat.[20] He in turn drew his inspiration largely from Vauban and the whole tribe of self-important French engineers who had followed.

The trouble with the engineers of the mid-nineteenth-century French service – and hence by imitation also those of the American – was that although they represented only one specialised branch in the art of war they were permitted to posture and parade as experts in the whole. They knew only too well that they were the most highly educated members of the officer corps, and they jealously guarded their status as spokesmen on any issue of strategy or military organisation. In France this meant that they worked against any moves to set up a true general staff, since such a body would have relegated them to the backwater in which they rightly belonged. It also meant that they did nothing to meet demands for systematic war planning against Prussia: instead, they merely repaired some parts of Vauban's ageing fortress chain and added a few new works to it.[21]

In America before the Civil War the situation had been a little different, although at heart perhaps not all that much. Admittedly, there was no remotely serious military threat, but there was an extensive chain of coastal fortifications to be maintained, and hence a strong engineer vested interest to be defended. There was also widespread apathy about creating a general staff.

Unlike some of their more reactionary French counterparts, Mahan and his disciples – most notably Halleck and Beauregard – did know and quote from the strategic writings of Jomini. They also drew enormous inspiration from Napoleon's *Maxims* – albeit probably less for their author's skill with the pen than with the sword.[22] These West Point writers were abreast of modern concepts of mobile warfare as perhaps the French fortress inspectors were not. Nevertheless, the Americans remained indelibly loyal to their own calling as engineers, and took every opportunity they could to rewrite Napoleon and Jomini in the direction of digging in and moving earth. The resulting doctrines contain Delphic paradoxes which might have been richly appreciated in the École Polytechnique, but which must have left the American citizenry-in-arms somewhat confused. They seemed to be saying that armies should move quickly at the same time as fortifying themselves; should travel light yet carry all the apparatus necessary for siege warfare; should be reckless and yet terribly, painfully, careful. Just as George Orwell's anthropomorphic pig in *Animal Farm* used sleight of hand to teach the animals to reverse their hard-learned slogans, so Mahan and his school seemed to be preaching that 'the rifle and bayonet are good – but the rifle, bayonet *and spade* are better'.[23]

This critique failed to examine the nineteenth-century record in detail, since in Napoleonic and Crimean battles – not to mention Bunker Hill and the Alamo itself – the side with the fieldworks had usually been defeated by the attacker. In 1813–14 the French had gilded the lily of their towering mountain defences on the Pyrenees by adding redoubts, but these had quickly fallen to Wellington's assault. Even at New Orleans, when a five-foot-high earthwork apparently stopped two British assault brigades dead in their tracks with well over 50 per cent casualties, we still have much room to speculate that a quite different result might have occurred if the attack had been properly managed, and sent in at a single impetus on the first day.[24]

It would seem that Mahan, Halleck and Beauregard based their faith in fieldworks less upon a careful study of history than upon their ingrained prejudices as engineers and also (perhaps equally important) as professionals. They certainly had no faith in the ability of militia armies to hold their own in the complex manoeuvres of regular warfare,[25] and they were only too fully aware of the appalling fragility of improvised American armies such as Andrew Jackson's at New Orleans. Hence they believed that, as far as America was concerned, the techniques of advanced modern tactics were available only to the 16,000 long-service regulars. These might perhaps be 'stretched' a little, as they had been for the Mexican invasion of 1846, but for any large-scale mobilisation their benign influence would be totally dissipated. The war would inevitably be fought by militias – and that meant it would have to be fought by primitive tactics which sacrificed mobility and flexibility in order to give a minimum standard of confidence and security to the troops.

For Napoleon the solution to the problem had been to stiffen raw levies with exceptionally large numbers of cannon, as well as with veteran *cadres* for training and indoctrination. For Mahan and his school, however, the answer lay in fieldworks:

Habits of strict obedience, and of simultaneous and united action, are indispensable to carry out what the higher principles of the military profession require. New and undisciplined forces are often confounded at the evolutions and strategic and tactical combinations of a regular army, and lose all confidence in their leaders and in themselves. But, when placed behind a breastwork, they even overrate their security. They can then coolly look upon the approaching columns, and, unmoved by glittering armor and bristling bayonets, will exert all their skill in the use of their weapons. The superior accuracy of aim which the American has obtained by practice from his early youth, has enabled our militia to gain, under the protection of military works, victories as brilliant as the most veteran troops.[26]

We have already discussed the myth of the American marskman. The much bigger myth which is also implicit in this engineer doctrine is that American militia armies would never be able to fight by manoeuvres in the European manner. Simply because they were militia, it was assumed that they would be forever stuck at the abysmal level of the 1815 Louisiana volunteers; that they would all inevitably be 'three month men' and incapable of improvement by long-term training or experience. There was in fact considerable evidence to support this assumption during the hectic spring of 1861, but from then on the armies gradually matured into something considerably better than the pre-war engineer school would have imagined was possible.

The Civil War may perhaps be seen as a struggle between two opposing doctrines of tactics. On one side there were the tactics of the drill manuals – translations from the French that had been lifted somewhat out of context and left unexplained, but nevertheless studiously learned by rote. These were the modern tactics of manoeuvre and decisive battle – the fruit of fifty years' cogitation on the lessons of Napoleonic combat. There is evidence that many American officers understood and sympathised with the basic assumptions of these tactics, but unfortunately none of them took the trouble to write books explaining the background, or adapting the drills to the particular learning processes which the Civil War armies would face. The result was that these advanced methods were often badly applied on the battlefield. They failed to be totally convincing in the American context – mainly due to faulty indoctrination – even after the armies had developed the skills necessary to handle them.

On the other hand the engineers could offer a quite different set of tactics – a 'kindergarten' set designed specifically for use by shambling armed mobs – which enjoyed an extensive background literature and doctrinal explanation. This was the idea of fieldworks, which, however much it may have been spiced with lip-service to mobility and manoeuvre, represented essentially a static conception of tactics. It was superficially attractive to the Civil War armies and purported to meet their needs as 'militia' troops. It was authoritatively recommended by the leading West Point experts and naturally appealed to those who were suspicious of European complexities and obscurantism. Its simplicity and its promise of personal safety for the individual gave it a head start over the risky choreography of Hardee and Scott – and by the end of the war it had generally been accepted in practice as the best way to fight.

THE PSYCHOLOGICAL POWER OF FORTIFICATIONS

There can be no doubt that a high combat value was attributed to fieldworks by many Civil War soldiers in almost every phase and theatre of the conflict. We have seen generals digging in nervously in 1862 no less than in 1864, and others just as systematically refusing to attack entrenchments during the same campaigns. As early as First Manassas the Union soldiery could be bewitched by talk of mythical enemy 'masked batteries', while the Confederate commanders were deterred from pursuit by fearsome reports of the defences of Washington. Grant aft r Shiloh and Lee after Antietam were each bitterly reproached for having failed to dig in, whereas the platonic minuets of Sherman and Johnston in the fieldworks north of Atlanta have drawn ecstatic praise from most critics. For many contemporaries it was a willingness to entrench that represented the touchstone of military prowess in the 1860s, and it is perhaps no accident that one of the most persistent anecdotes of the war is the one about the soldier taken to task for failing to clean his rifle. 'I know my gun is dirty, Colonel,' comes his reply, 'but I've got the brightest shovel you ever saw.' [27]

In his perceptive study of these problems Edward Hagerman showed that the importance accorded to fortification was less a reaction to new weapons appearing on the battlefield than a product of ingrained, doctrinaire book-learning. Halleck and McClellan did not dig themselves in as a response to the 1862 Confederate armament of shotguns and bowie knives, but simply because they felt it was the 'scientific' way to fight. By the same token the widespread psychological fear of attacking fortifications later in the war had nothing to do with some new prowess of riflemen firing at long range, but everything to do with a combination of exaggerated fears, doctrinal orthodoxy, and war-weariness.

Fieldworks could consist of a number of different elements, some designed to prolong an attacker's approach under outgoing fire, others to protect the defender's troops from incoming fire. Abattis, palissades and ditches were used to impose delays upon an assault at a range where extra casualties could be caused by defending musketry. This range would rarely be more than a few yards in the case of a ditch – because the defender would have to see into it to be effective – or it might perhaps extend to twenty or a hundred yards in the case of timber entanglements. In either case, however, the trap would be equally deadly whether smoothbores or rifle muskets were being used.

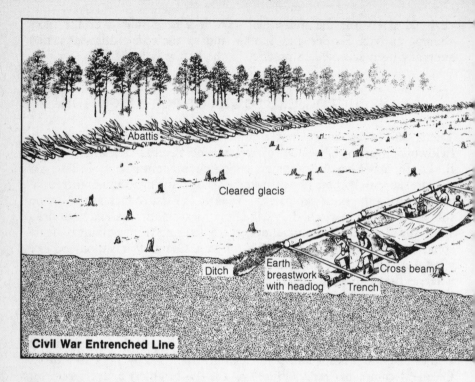

Abattis

Cleared glacis

Ditch

Earth breastwork with headlog

Cross beam

Trench

Civil War Entrenched Line

Those elements of fortification designed to give protection to the defender might consist of a trench with a parapet of spoil thrown up to the front and a headlog placed along the top. If there was time, loopholes could be dug below the headlog and crossbeams inserted to make a frame against collapses or a shelter against the weather. In essence, however, the idea was simply to allow the defender to shoot without exposing more than a very small part of his anatomy to the enemy's view. There was nothing particularly complicated about this requirement, even though some of the more lovingly crafted examples of the art must have made it seem so.[28]

Pens improvised simply from fence-rails or planks were sometimes used with effect. They could give protection against musketry although not against artillery – as the Confederates found to their cost at both Spotsylvania and New Hope Church.[29] But as an 'instant breastwork' which could be constructed quickly or even carried into action by the assault troops themselves solid planks or rails were invaluable.[30]

It may be worth pointing out that the physical attributes neither of obstacles nor of trenches would usually be enough in themselves to stop an attack. Only the very deepest ditches and highest parapets, bordering on the category of 'permanent fortification',[31] could hope to do that. Most abattis could be crossed quite quickly, most ditches and ramparts scaled, and most rifle pits invaded by a determined assailant.[32] All they could really do for a defender was to win him extra time to fire more shots from relative security before the attack or its firepower started to bite. The hope was that this prolongation of defending fire – all delivered at short range – would outface and demoralise the attacker before he reached his objective.

Obviously a defender standing in the open, with neither obstacles to his front nor cover for himself, would be considerably more vulnerable than a defender who enjoyed either (or preferably both) of those advantages. Nevertheless, one wonders if there was really very much physical difference between the status of a properly fortified unit and one which was unfortified but skilfully lying prone behind ploughed furrows or a dip in the ground, with an open field of fire to its front. Such a unit would not present a very great target to the enemy, yet could still bring heavy musketry to bear on the vital last hundred yards. Admittedly it would lack some of the freedom to move arms and bodies which a trench might provide – but it might also feel less isolated than in prepared positions, since these tended to be rather lightly garrisoned.

Estimates varied on the optimum density of manpower for a defensive line, but 'Johnny Green of the Orphan Brigade' was certain that the total (including reserves) of one man per yard or one per two yards had been inadequate in his battles at Chattanooga, New Hope Church and Jonesboro.[33] Colonel Nisbet of 66th Georgia in the same campaign wanted between 2·5 and 5 men per yard, and General Sherman recommended between 1·8 and 2·5 (that is, between 3,000 and 5,000 per mile).[34] Even the lower figures quoted as acceptable in these Western battles turn out to be quite generous when compared with the Confederate manpower available for the trenches at Petersburg, yet they are all well below the densities seen in most Eastern battles fought on open ground.[35]

Actually the main physical strength of a trench position was usually to be found neither in the extra protection it offered the defender nor in the obstacles it put in the way of an attacker. Paradoxically, it was the cleared field of fire in front of the trench that made it most dangerous. In broken country such an open glacis would not normally be available at convenient points in the line, so it made all the difference if troops had found time to clear one artificially before battle was joined.[36] It gave them a killing ground in which an attacker could be brought face to face with the full dangers of his enterprise. Yet – at the risk of repeating ourselves – we must remember that a cleared field of this type was of importance only at short range, not long. Even when entrenchments were constructed with long fields of fire – itself less common than one might suppose – there seems to have been little attempt to place markers to help range estimation. This might sometimes be done for the artillery,[37] but it was very rare for infantry. The effect of musketry was perceived as applying mainly to the last hundred yards, making it largely irrelevant whether smoothbores or rifles were used. Provided that he could bring an adequate number of either into action at the correct moment, a defender could hope to draw the full benefit from his preparations.

Just what could this benefit be? Many Civil War soldiers imagined it as almost unlimited in scope, especially since Halleck had estimated in the 1850s that 'a judicious system of fortifications' multiplied the soldier's combat value by no less than six times.[38] Admittedly, he was thinking more of permanent fortifications than of fieldworks, but the alluring prospect of 10,000 men beating off 60,000 was too good to be ignored and was sometimes mentally transferred to more transitory types of entrenchment. General Jacob D. Cox stated that 'one rifle in the trench was worth five in front of it',[39] and a member of 14th Indiana was still more positive when he stated:

I have seen strongholds stormed, and I know that if brave men defend a redoubt it is physically impossible for flesh and blood to oppose solid earthworks and solid shot.[40]

According to this calculus, presumably, the entrenched army of 10,000 brave men could look with equanimity at an attack mounted in any numerical strength at all.

Such extravagant claims are certainly in keeping with the sheer scale of digging that was undertaken. The Washington defences which had outfaced Johnston in 1861 were 37 miles in length,[41] while those built by Lee at Richmond in the same year (thereby earning him the nickname 'Ace of Spades')[42] were of a comparable order of magnitude. Thus it was not only towards the end of the war that such great works were undertaken, although the honeycomb of trenches, redoubts, dug-out shelters and rifle pits which later appeared before Vicksburg, Petersburg, Chattanooga and Atlanta did indeed eventually come to outclass everything that had gone before. One Union officer returning to Chattanooga in 1864 expressed his awe as follows:

A veteran European soldier would be amazed should he travel over the route of our army from here to Atlanta – a hundred and thirty miles through a constant series of defences, mountains and rivers supported by the labours of thousands for months.[43]

As far as the tactical effect of the sophisticated late-war entrenchment is concerned, the general consensus was that 'It seems impossible to charge and take such works, yet sometimes our boys do it, but the greater part of them are taken by General Sherman's flank movements'.[44] Earlier in the war there had perhaps been slightly more optimism about the chances of carrying fieldworks, but in general terms the picture is consistent throughout: experienced soldiers try not to attack entrenched positions if they can possibly help it. It is true that there was usually a supply of inexperienced troops willing to try their luck against fieldworks and a rather greater supply of officers, both experienced and otherwise, to help egg them on.

Sometimes their assaults were successful and sometimes they were not, but always they suffered heavy casualties, which must have helped deter them from attempting the experiment a second time.

The longer the war went on, the more soldiers could be found who had experienced a 'slaughter pen' at first hand. Such men had searing visions of the human cost of such enterprises, and quite naturally found it difficult to balance this against the highly abstract benefits to be gained by even a successful assault:

An attack on fortified lines must cost a fearful price, and should be well weighed whether the cost exceed not the gain. This, then, is what the assault means – a slaughter-pen, a charnel-house, and an army of weeping mothers and sisters at home. It is inevitable. When an assault is successful, it is to be hoped that the public gain may warrant the loss of life requisite ...[45]

Another veteran was even more trenchant on the subject of failed assaults:

It was certainly no evidence of capacity, or even courage in an officer, to recklessly lead his command into a slaughter-pen – and all such desperate charges as Marye's Hill, and Bloody Angle and Kenesaw told in verse and story as evidences of what American soldiers can do and dare, had no appreciable effect, except to prolong the war.[46]

Both of these quotations come from Sherman's army in 1864 – the epicentre of respect for fortifications and dislike of assaults. The second of the two commentators had been present at Kenesaw Mountain, and was later to demonstrate his belief in the live-and-let-live system by contriving a firefight which lasted all morning and cost his regiment one man wounded. After this, since the day was hot, he led his troops to the river separating the two armies and they bathed amicably with the Confederates.[47]

By 1864 it seems that there were numerous cases of combat refusal when an attack on fortifications was proposed. Even if this did not lead to a formal mutiny it could often lead to an 'attack' which went to ground almost before it had crossed its start line. The abortive 'battle' of Mine Run was an example of this on the grandest possible scale, since the entire Army of the Potomac came into position to storm Lee's works but then thought better of it and went home.

The action of 35th Massachusetts at Spotsylvania provides a good example of how the 1864 fighting must often have been conducted. The regiment advanced behind another unit until it came under fire, when a bounty-jumper shouted 'Retreat!' and the whole regiment routed in panic. They rallied calmly when they had regained their own earthworks, insulted their new general (whom they did not recognise), then advanced again to a position in open ground one hundred yards from the enemy entrenchments. Then the whole line lay down, without firing a shot, and in this position calmly sustained the fire of the enemy two or three hours, with little loss to us as the shells and bullets of the Confederates passed over our heads. The order was simply to 'feel the enemy', and as it was plain they were ready to receive us, no final assault was ordered.[48]

The regiment lost a total of twenty-two casualties in the course of the entire day's fighting, which is astonishingly low for a unit which had been so near the enemy for so long. But it is clear that 35th Massachusetts had by this time grown weary of taking risks. Its historian tells us that all the best men had now gone, leaving only the less confident soldiers and the bounty-jumpers.[49]

The experience of 35th Massachusetts was by no means exceptional. Perry Jamieson lists an impressive series of combat refusals and half-hearted attacks from Spotsylvania, Cold Harbor, Petersburg and elsewhere,[50] which are perhaps best summed up in the words of a Confederate colonel at the battle of Jonesboro: 'The men seemed possessed of some great horror of charging breastworks which no power, persuasion, or example could dispel . . .'.[51] The impression we get is that by 1864 the mere existence of a fortification, however technically weak it might be, would usually be enough to forestall a serious attack. Digging a trench was thus in part a symbolic staking of a claim – a signal to the opponent that this part of the battlefield was no longer negotiable tender.[52] That in turn led to an escalating spiral of ritualistic trench-building, whereby each side tried to move as much earth as it could in order to lay claim to as much real estate as possible. The armies learned to dig in automatically wherever they halted, as much to warn the enemy not to try anything foolish as to give themselves physical protection from shot and shell.

An important part of this was perhaps the outward appearance of the trench itself, regardless of its actual defensive value. The enemy would be shown 'a yellow line of dirt'[53] where the parapet had been thrown up, so he would be left in no doubt that something serious had been done to improve the combat power of the landscape. Yet at the same time the defending soldiers would be invisible – and sometimes effectively camouflaged behind shrubbery or abattis. Because they could not be seen, the defenders would possess a heightened aura of menace for the attackers. They would be less easy to relate to on a personal level – more inhuman. Thus the simple process of turning a little earth would have achieved an effect not dissimilar to that of the classic 'terror weapons' of history.

A logical theorem may be drawn to show that if both sides in a war believe that it is impossible to storm breastworks then neither side will feel safe until it is sitting behind them, and no one will wish to put a serious effort into storming the enemy's. Hence both sides will actually have gained security, regardless of any material weaknesses in the trenches themselves, because their beliefs will have turned into a self-fulfilling prophecy. The more strongly those beliefs are held, furthermore, the greater will be the mutual reinforcement and hence the greater military security for the two armies.

It is probably in civil wars that this process of mutual doctrinal reinforcement can operate most fully, since the two armies share a common heritage and language. Their ideas can flow more easily from one side of the battle line to the other, especially if all the generals have been educated at the same school. If there are no active central staffs concerned with formulating doctrine, moreover – as there were not in the Americas of the 1860s – there will be no driving force to shake the armies out of their ingrained habits of thought. Admittedly, a number of attempts were made to apply an opposite theory of tactics in the Civil War – the *attaque à outrance* (and the Confederates were certainly able to apply such a system more effectively than the Federals) – yet the fact remains that by 1864 the great majority of armies and commanders, gray as well as blue, had wholeheartedly embraced the ideology of the engineers.

All this poses a problem of evidence for the historian, since at this distance in time it is difficult to disentangle the psychological power of entrenchments from their physical advantages. We can certainly say that on many occasions trenches were dug when they need not have been, and that on many more occasions they outfaced attacks when by rights they should have been carried. They were often used as an excuse to avoid combat by battle-weary soldiers who had been driven too far, and they were often the product of a fleeting military fashion rather than 'the outgrowth of the intelligence of the American volunteer applied to the experience of many bloody battles'.[54] They may have saved some lives in the short term, but they also helped to defer military decisions – and thereby made the war longer and costlier than it might otherwise have been.

Yet against this view we have repeated assertions that the rise of trenchworks in the Civil War was the inevitable result of improved weapons and a new depth of understanding of modern combat; that the largely unfortified Napoleonic battlefield was forced by *physical* circumstances to give way to the underground burrowings of later wars. We read of green units disdaining cover in their first combat, but learning by hard experience that it was essential in all their later actions.[55] For example, at Kernstown in 1862 F Company, 21st Virginia, was tempted to rise from the shelter of a fence to repel an attack:

The men along the fence left its protection and fought as I never saw any fighting during the war. After this, they were glad to take advantage of anything.[56]

We even find that whole armies congratulated themselves on their skill in learning about fieldworks – for example, Sherman's men mocked Meade's for the heavier casualties the latter sustained in frontal assaults. They could point to their own 31,687 casualties in the period May–June 1864, as against his 88,387, to show the significant material benefits which a respect for the spade could bestow. In reply the Eastern soldiers could only mutter that these same statistics proved who had fought the harder.[57]

There is probably some truth in the idea that the use of earthworks was beneficial to any army's tactics, in the same way that Wellington's use of reverse slopes and regiments lying prone helped him to win his battles. There was obviously no advantage in standing in the open under fire to be 'damnably mauled' like the Prussians at Ligny; and we should not be tempted to follow the Northern newspaper which wanted to 'dismiss from the Army every man who knows how to build a fortification'.[58] Nor should we automatically accept the full strictures of General Hood against the demoralising effect of the entrenching habit, even though they were to be echoed quite closely by Sheridan and even Sherman himself.[59] But we must nevertheless be on our guard against excessive or anachronistic claims for what earthworks could and could not do.

Earthworks could not confer any special mystical advantages upon the American soldier which were denied to others, or bring out latent skills in marksmanship which he had not previously possessed. They might have been especially attractive to Americans as a result of specific national mythologies and military training curricula, but that was a matter of taste rather than of hard military logic. Only rarely could they enhance the natural protection of the terrain, because in the Civil War there was usually enough protection provided by nature already. It needed but little improvement to create a stronghold, and that was more often a matter of levelling a field of fire than of placing obstacles or shelters on already level ground. It is significant that most of the great 1864 trenchworks were built in forests.

Nor were deep trenches needed to protect men from sustained heavy shelling at long range, as they were to be in later wars, since such shell fire had not yet appeared in battle. The threat was mainly from direct fire, and hence a shallow scrape could often serve as well as an elaborate dug-out.

It is hard to find any evidence at all to support the notion that the elaborate earthworks of the Civil War were any more necessary than they had been in Napoleonic times. There were admittedly some marginal new threats from snipers and rifled artillery which may have exerted a dis-proportionately great psychological terror upon troops accustomed to more localised weapons, and at the start of the war there was some justified caution about exposing raw militia to excessive dangers. Yet by the middle of 1862 it must have been clear that the new armies would fight with gusto after all, and that the new weapons were less different from their predecessors than had been claimed. In these circumstances the reasons for the change from Napoleonic practice must be sought almost entirely in the minds of the combatants. A more educated American population was less ready to risk death without at least a semblance of personal protection, and a high command imbued with the flannelling of the Vauban and Mahan schools was blinded to the inner character of mobile warfare. Once this curious brew had been mixed together and shaken up thoroughly in a few pitched battles, it settled out as the 1864 elixir. Lots of digging, lots of skirmishing, noise and smoke,[60] lots of respect for the enemy's line and an acute awareness of the claims he had staked. But not often very much real fighting. It was a far cry indeed from the methods of Napoleon!

6

The Infantry Firefight

The characteristic mode of combat in the Civil War was the infantry firefight at close range. This was the usual result when an attacker had overcome his first fear of the enemy's rifles or fieldworks and had marched into contact. Sometimes a decision could be reached before this by long-range fire, and very occasionally it could be reached entirely by cold steel. More normally, however, there were clear winners and losers only after a period of intense musketry.

TWO THEORIES OF INFANTRY FIGHTING

We have seen that one very influential Civil War theory – in fact the most influential of all – envisaged infantry sheltering behind fieldworks, delaying the attack by obstacles and discouraging it by heavy fire. According to this theory the contest would almost inevitably be won by the static defender, because he would enjoy the advantage of protection while the attacker was presented as an easy target in the open.

We have also seen that by 1864 troops imbued with this theory had become reluctant to press home their attacks, since they understood only too clearly what dangers they would have to run. Perhaps less obvious, however, are some of the purely logical difficulties for an attacker which such respect for entrenchments would entail. For example, the attacker who believes in fieldworks will naturally have built some of his own – and these will create a physical obstacle to his advance no less than a psychological reminder of the safety he is leaving. On some occasions soldiers were expressly forbidden to charge forward from their works even in favourable circumstances.[1] Then if the attacker does come close to the enemy he will perceive the latter's advantage in terms of protection from fire. Instead of moving forward, the attacker will himself wish to gain

137

protection by going to ground and even digging in. Hence the contest will subtly have changed in structure from being an 'attack' against a 'defence' into a parallel contest of two entrenched units. Deprived of its dynamic, the battle will degenerate into an exercise in attrition and exhaustion.

Sometimes an attack would go forward, find it could make no progress, then retire a little to a somewhat safer place – at perhaps 300-400 yards from the enemy – and entrench.[2] On occasion some natural cover might be found closer than that, as in the Iron Brigade attack at Antietam already mentioned:

After a few rods of advance, the line stopped by a common impulse, fell back to the edge of the corn and lay down on the ground behind a low rail fence.[3]

This attack eventually moved forward after reinforcements arrived, but was then repelled in turn by the arrival of Confederate reserves.

Another classic case of natural cover being too tempting to resist came at Fredericksburg during the main Union assault on Marye's heights. The attackers lay down behind a swell in the ground 75–300 yards from the enemy, and maintained their fire from there. As additional regiments arrived, they simply joined in the firing, advancing no further.[4] Some men exfiltrated as individuals during the remainder of the day, but most whole units waited for nightfall – or even the second night – before they gave up the contest. At Kenesaw Mountain the attack dug in even closer to the enemy and stayed longer: thirty yards and three days respectively.[5]

One psychological reason for digging in so close to danger lay in the still greater danger, once one had come so close, which was assumed to lie in retreating back across open ground.[6] When one of the veterans of the Atlanta campaign described the difficulties and losses involved in coming close to the enemy, he culminated with the following snort of disgust:

... And then the slaughter of a retreat *there*! Oftentimes it is preferable to lie down and take the fire there until night rather than lose all by falling back under such circumstances.[7]

Perhaps this fear of retreat was linked to a horror of turning one's back on the threat. Even though the real danger from fire to a unit retreating from 200 to 400 yards from the enemy may not actually have been terribly great, it must have seemed crushing to a unit which had already suffered much while facing forward. A type of reverse ostrich syndrome may have applied, whereby the danger was bearable only while the men continued to watch it. Reluctance to abandon the wounded may also have played a part, as indeed may justifiable possessiveness for the ground won at such cost.

For whatever reason, Civil War units on the offensive did often settle down to a prolonged exchange of shots at close range with a defender, even without fieldworks being dug by either side. The following case from Second Manassas is characteristic:

... it was a stand-up combat, dogged and unflinching, in a field almost bare. There were no wounds from spent balls; the confronting lines looked into each other's faces at deadly range, less than one hundred yards apart, and they stood as immovable as the painted heroes in a battle-piece ... and although they could not advance, they would not retire. There was some discipline in this, but there was much more of true valor.

In this fight there was no manoeuvering, and very little tactics – it was a question of endurance, and both endured.[8]

This combat continued from 'late in the afternoon' until at least 9 p.m. – perhaps three or four hours of close musketry. Another example comes from Murfreesboro:

A few of the more venturesome Rebels reached a rail fence less than twenty yards from the muzzles of our muskets, but none ever returned. Others paused at a distance of seventy-five to a hundred yards, delivered their fire and dropped to the ground to load and fire again ...
For at least half an hour this bloody angle at the edge of the cedars was held ...[9]

A similar battle is recorded by one of the Union attackers at Champion's Hill in the Vicksburg campaign:

The enemy had fallen back a few rods, forming a solid line parallel to our own, and now commenced in good earnest the fighting of the day. For half an hour we poured the hot lead into each others' faces. We had forty rounds each in our cartridge boxes, and probably nine-tenths of them were fired in that half hour.[10]

On the second day of the Battle of Seven Pines there was an interesting series of actions which have been recorded in detail by General Gustavus Smith.[11] Some fifteen Union regiments under General Richardson appear to have been involved in firefights lasting about an hour and a half, with reliefs sometimes replacing them or their ammunition. In some cases there was a resolute charge which cleared the front for a time; in other cases there was a static battle. The normal experience seems to have been that the Union forces advanced slowly and pushed back the Confederates a short way, firing as they came. This was woodland combat and the fire was 'very close and deadly',[12] sometimes between two friendly units which mistook each other's identity in the undergrowth. Richardson's total loss was some 1,050 casualties, or an average of 70 men per regiment – rather less than one man per regiment per minute of the firing. The general impression of how the battle developed is confirmed by General Hooker, who was engaged on a supporting axis at the same time:

From the beginning of the action our advance on the rebels along the whole line was slow, but I could feel that it was positive and unyielding. Our lines were well preserved, the fire brisk and unerring, and our troops reliant – all the omens of success.[13]

Hooker's loss was 153 casualties in seven regiments, or 22 per regiment – one man hit per regiment every four minutes. The overall Confederate loss was at least 800 men from up to eighteen regiments, making perhaps 42 casualties per regiment, or one per regiment every two minutes.[14]

None of this adds up to firefights which were concluded rapidly, although it could sometimes happen. What we see more usually is a fairly protracted period of firing with only a limited forward movement by the attacker, if there is any at all. Maybe a conclusion could be reached in thirty or even fifteen minutes; maybe it took the $1\frac{1}{2}$–2 hours required for a regiment to fire off all its ammunition; or maybe it would drag on until night – however long that might be. The timespan was very variable, but the central thought behind many of these contests was that victory depended upon firepower, that the way to win was to blaze away into the enemy's face until he was too numb, or too exhausted, or too badly mauled to carry on. Then perhaps he would go away. This was the hope which kept men going in many of the assaults and slaughter pens of the first three years of war, and sometimes even in the year of demoralisation itself, 1864.

It was perhaps foolish for attackers to imagine that they could use firepower and protection more effectively than a prepared defender, and to that extent the eventual disillusionment with assaults was justified. Yet in battle after battle we find regiments hanging on at close range, slugging it out with presumably this hope in their minds. Perhaps it was a matter of pride in the imagined prowess of the unit, or simply fear of the disgrace of withdrawal. Probably there was a complex mixture of psychological pressures generated by the tensions of close combat itself. But, whatever the subliminal reasons may have been, the only rational theory of tactics that could be applied to these firefights was the static, attritional one of the engineers. The idea was to exchange shots and make both sides bleed, until the side with more men would be left the winner.

The alternative theory of tactics was the one implied by the European drill books, which may be termed the 'infantry' as opposed to the 'engineer' solution to the problem. According to this theory it was as ridiculous for a defender to stay cowering behind his earthworks as it was for an attacker to go to ground at the very moment when he had finally come near his objective. What was required instead was a fluid battle of attack and counter-attack in which firefights would be very short indeed, if they took place at all. The decision would not depend upon firepower and protection, according to this theory, but upon the bayonet and shock action.

'Shock action' is a much misunderstood term in the Civil War context, and the bayonet had been mercilessly mocked ever since General Sherman first realised that his army was shy of Johnston's trenchwork. In the high-handed verdict of history these traditional instruments have been derided as representing a theory of combat which no longer worked in practice, a theory relegated to the scrap heap by the supposedly more 'scientific' conceptions of the engineers. Yet if we examine them here in a little detail we may perhaps discover that they were not quite as irrational or inappropriate as the polemicists would have us believe.

The key word, for both attack and defence, was not so much 'bayonet' as 'shock'. The defender would hide himself as best he could until the enemy was close, then spring up, shock him and charge into him, chasing him away. Conversely the attacker would attempt to come close to the defender, shock him into flight and then – once again – chase after him. This is what the British in the Peninsular War had generally understood by infantry tactics, and it was also the concept adopted by many French writers of the early nineteenth century – notably Marshal Bugeaud, the great mentor of the army which eventually fought so well in the Crimea and Italy. Both he and Wellington had seen that the important thing was to keep up the forward impetus of a formed body, regardless of whether their shock effect was gained by the bayonet alone, or by cheering, or by both of those allied to a short, sharp volley. This concept was perfectly familiar to the teachers at West Point, and it had worked well in Mexico in 1846,[15] but the lack of theoretical explanations in the drill manuals meant that it was rarely properly understood by the hastily improvised officers of the Civil War. In the absence of systematic guidance from above they tended to follow the most obvious – and superficially safest – line of action, using musketry from a protected position.

A great deal of misunderstanding has arisen from the fact that a 'bayonet charge' could be highly effective even without any bayonet actually touching an enemy soldier, let alone killing him. One hundred per cent of the casualties might be caused by musketry, yet the bayonet could still be the instrument of victory. This was because its purpose was not to kill soldiers but to disorganise regiments and win ground. It was the flourish of the bayonet and the determination in the eyes of its owner that on some occasions produced shock,[16] just as on other occasions the same effect might be produced by noise or muzzle flash. It is only the believers in the attritional 'engineer' or 'fire-power' theories of tactics who have tried to portray the bayonet as a killing machine, in order to ridicule it.

A crucial element in shock action was the holding of fire until the enemy was at very close range. Not only would one's first volley be physically more damaging, but it would come as more of a surprise, especially if the enemy himself was unable to return the compliment at once. When he was subjected to a mass attack at Antietam, General Gordon reckoned that

The only plan was to hold my fire until the advancing Federals were almost upon my lines. No troops with empty guns could withstand the shock. My men were at once directed to lie down upon the grass. Not a shot would be fired until my voice should be heard commanding 'Fire!'[17]

In the event this plan worked well; the volley was very destructive and the counter-charge which followed it was effective, although it failed to completely break the opposition. The battle settled down into a protracted firefight after all. Rather more successful was the Orphan Brigade's use of a similar tactic near Resaca: 'Our infantry reserved their fire until the enemy were in close range when we opened on them, almost every shot killing a man.'[18] After this the attack made no further headway, although the defenders felt too weak to counter-attack it.

The use of shock on the defensive often exploited the classic 'ambush' tactics of concealment and surprise. Shock tactics could also be used in the assault, however, without the benefit of concealment. A fortnight after its Resaca action the Orphan Brigade had to attack a hill. It was told to 'Reserve your fire until forty yards of their line, then fire, each man taking careful aim and then rush upon them and give them the bayonet'.[19] This was successful and, despite early Union shooting at longer range, the Confederates delivered their own fire only when they could see their enemies' buttons. The Federals broke as the attack came to ten yards of them. Then a number of them were shot in the back.

Nor was any of this new. As early as the Peninsula campaign of 1862 there had already been a fairly widespread use of shock tactics, especially in the Confederate army. At Gaines's Mill A.P. Hill's troops had stalled before the strong Union position when Whiting's division was fed through them in one of the war's more successful examples of the passage of lines. General Law tells us that his brigade was told not to halt at that point, but to take up the charge

in double-quick time, with trailed arms and without firing. Had these orders not been strictly obeyed the assault would have been a failure ... [A thousand men were hit as they advanced, but] not a gun was fired in reply; there was no confusion, and not a step faltered as the two gray lines swept swiftly and silently on; the pace became more rapid every moment; when the men were within thirty yards of the ravine, and could see the desperate nature of the work in hand, a wild yell answered the roar of the Federal musketry and they rushed for the works. The Confederates were within ten paces of them when the Federals in the front line broke cover, and leaving their log breastworks, swarmed up the hill in their rear, carrying away their second line with them in their rout.[20]

After this, once again, the main Confederate fire action came while the enemy was fleeing from them.

Gaines's Mill may not actually have been won with quite the effortless virtuosity portrayed by Law, but it was certainly won at the point of the bayonet by troops who knew how to hold their fire in the attack. This was a lesson particularly taken to heart by John B. Hood, who commanded the other brigade in the decisive movement. When 150 yards short of the enemy line he is reported to have told his men: 'Don't halt here! Forward! Forward! Charge right down on them and drive them out with the bayonet!'[21] This was to be the hallmark of his fighting style thereafter. Hood's main fear was that an attack might stall if the men let themselves be drawn into a firefight at long or medium range. The aim was to get close, break into the defender's carefully prepared firing line and then exploit the confusion which followed. Hood believed that once the first line had been pushed back the pursuit could be pressed hard so that no subsequent lines in the rear could hold out. He was quite specific that the purpose of tactics was not to kill enemy soldiers by fire but to disrupt and disperse enemy regiments by charging through and over them: 'Safety in time of battle consists in getting into close quarters with your enemy. Guns and colours are the only unerring indications of victory.' [22] There is certainly evidence to show that a unit which continued its charge without stopping could cause astonishment in the enemy's ranks, simply because such a tactic was so unexpected. Civil War soldiers regarded it as more natural to fight statically by fire, and did not know quite what to make of an attacker who held to some other theory. In Captain De Forest's first battle[23] his green troops started to receive infantry fire at about five hundred yards from the enemy trench. After marching forward through three volleys the colonel allowed them to relieve their feelings by opening fire and seeking cover on the ground, but then he insisted that they resumed their advance. This they did, still firing as they moved, although without regularity of drill. The enemy ran off when the range had closed to about one hundred yards. Prisoners later told the regiment that 'We expected to see you come on in the usual style, – halt to fire and then advance again – not fire and come on all together ... Your firing didn't hurt us ... but your coming on and yelling scared us.' [24] De Forest concluded that the position would not have been carried without both the firing in the advance and the rapidity of the advance itself. The firing raised friendly morale and spoiled the enemy's aim to the extent that only six men in the Union regiment were hit by musketry, while the speed of assault created the enemy's urgency to run away. A slower attack might have both inflicted and suffered more casualties – and it would have failed to carry the position because it would have lacked the essential ingredient of shock.

The essence of shock tactics was the deliberate acceptance of a higher risk in order to achieve a more decisive result. Holding fire until close range was psychologically difficult for most soldiers, because it left them longer in danger before doing anything in return, yet it was likely to be more effective than uncontrolled blazing away at long range. By the same token it took courage to continue an advance upon the enemy without alleviating the danger by lying down or halting at decently long range. Thus commanders who asked their men to use shock tactics were asking for an unnatural response to fear. It took exceptional leadership, or exceptionally high-quality soldiers, to make the system work.

General Chamberlain tried to explain shock tactics to his men on the Quaker Road in 1865 by telling them to 'Once more, try the steel. Hell for ten minutes and we are out of it.' [25] In rather similar vein General Taylor in

1862 had told his men not to 'dodge' the bullets but to press forward, because otherwise they would be under fire for an hour.[26] The prospect of shortening the duration of danger must certainly have been attractive, although it was probably the 'ten minutes of hell' that loomed larger in most men's thoughts.

A further difficulty with shock tactics was that they demanded more co-ordinated action than the 'engineer' tactics of firepower and protection. Even quite a small minority of soldiers opening fire prematurely could spoil the shock effect of a supposedly crushing volley, just as a minority of stragglers could make an attacking spearhead believe that it had been abandoned to its fate and was sticking its neck out too far. Sherman noted that this effect was particularly acute with skirmishers, who always had to contend with the loneliness of their open-order formation anyway. He said that even one or two men hanging back could bring a skirmish line's advance to a halt.[27] Nevertheless, the same phenomenon did also apply to the movements of a formed line. The impact was lost unless a regiment stuck together and moved with unanimity and common purpose, if not necessarily with drill-ground precision.

Unanimity of purpose could be spoilt by several different influences. In De Forest's battle, described above, it was almost spoilt by the in-experience of the troops. Only exceptionally firm leadership – and also probably the weakness of the enemy's position – carried them through. With excessively experienced troops, by contrast, there would be too many 'old lags' in the ranks for bold ventures to be practicable. Whenever genuine risks were to be run – as opposed to operations which seemed risky to a greenhorn but were routine to a veteran – the unanimity of the regiment would tend to operate in the direction of security and safety.

Then again, pure chance might upset the balance in even the most total surprise. Consider the case of the Federal regiment at Second Manassas which used immaculate shock tactics only to find that ill fortune had pitted it against a rather special opponent. This is how the battle looked to one of the Confederates:

We neared the Chinn house, when suddenly a long line of the enemy rose from behind an old stone wall and poured straight in our breasts a withering volley at point-blank distance. It was so unexpected, this attack, that it struck the long line of men like an electric shock. But for the intrepid coolness of the colonel, the 17th Virginia would have retired from the field in disorder. But his clear, ringing voice was heard, and the wavering line reformed. A rattling volley answered the foe, and for a minute or two the contest was fiercely waged . . .[28]

The ultimate outcome was that after a firefight lasting ten minutes at fifty yards' range the Union line withdrew – surely an undeservedly dis-appointing outcome to a perfect deadfall.

The simple fact is that shock tactics were always difficult to pull off successfully unless one's own troops were entirely reliable or the enemy's were markedly inferior. In Wellington's battles his soldiers had been dedicated professionals and the enemy had gradually fallen to pieces, in terms of his combat quality, the longer the war went on. In the Mexican War the American troops had enjoyed a considerable qualitative advantage over their opponents, as had the French in Algeria. In the French battles of the 1850s and the American battles of the 1860s, however, the standards of drill did not always reach the level necessary to achieve shock, nor did the enemy always run away when subjected to it. In the Civil War, especially, there was such a mixture of units at different phases of their 'learning curve', on both sides, that it was impossible for a higher commander to make confident generalisations about the likely efficacy of any particular system of tactics. Thus Hood as a brigade commander could predict the efficacy of his unstoppable Texans, but Hood as an army commander was proved to be out of touch with reality when he applied the same standards to all his men. Civil War commanders found that each regiment had to be allowed to operate in the way to which it had become accustomed, in the hope that it would encounter the type of enemy it could 'whip' – or at least 'drive'.[29] Shock tactics promised the biggest dividends if they could be made to work, but fire and protection tactics seemed to offer the greatest safety – even if they did not always deliver it in practice.

MUSKETRY RANGES

An important feature of Civil War infantry fighting was the range of firing – both the range at which fire was first opened and the range at which an enemy could be persuaded to run away. These were rarely the same, although they might sometimes coincide – as we have seen – when shock tactics were being used. More normally an attacker would continue to advance some distance under fire before a decision was reached: he would observe a distinction in his mind between 'maximum' and 'decisive' range.

We have already seen many examples of protracted fire at very short range, and the conclusion appears to be unavoidable that this was the normal state of affairs in the Civil War.[30] We may nevertheless hope to refine our findings in a number of respects, since not all Civil War battles were the same. The terrain and the fields of fire were longer in some battles than in others, while the quality of the weapons improved quite significantly as the war went on. Early willingness to storm fortifications had sagged dramatically by 1864, and the incidence of strong fieldworks had correspondingly increased. There was also a difference in doctrine from one unit to another, from one commander to another, and from one theatre to another.

There is a major problem of evidence in all this, since armies unaccustomed to measuring ranges accurately, even in their prepared defensive positions, can scarcely be expected to cite accurate distances for confused battles which often took place on unfamiliar ground. Nevertheless, the literature is full of positive statements of the ranges at which fire was exchanged, so it may be worth taking them on trust. Even if this tells us no more than the ranges at which Civil War soldiers would in retrospect have *liked* fire to have been applied, it still represents knowledge of a sort.

In Gustavus Smith's book on Seven Pines there are seventeen usable references to the range of musketry fire, although unfortunately only three of them make the important distinction between 'maximum' and 'decisive' range (respectively 100 and 40 yards, 50 and 10 yards, 300 and 30 yards). The other ranges quoted are presumably those at which the bulk of each firefight took place, without necessarily telling us the range at which fire was opened or the range at which a decision was reached. In a good number of cases, however, it is specified that this was the range at which fire was opened.

The average range for the seventeen fights was 68 yards, or very considerably less than the 200 yards quoted by British experts as the battle range of the Brown Bess smoothbore musket.[31] There is only one reference in the Seven Pines sample to musketry at more than 200 yards, and that is the 300 yards maximum already mentioned. There is one reference to a firefight at 150–200 yards, but all the rest are described as taking place at 100 yards or less, with three at 100 yards, one at 75 yards and all the rest 50 or less. Twenty or thirty yards appears to be the most common combat range in this particular battle.

Seven Pines was not, of course, a 'typical' Civil War action. It had rather more close country than was normal in Virginia – although it did also include some open fields. It was very much an 'early war' battle, with a high proportion of smoothbore muskets and almost universally inexperienced regiments, yet it could show at least its fair share of fortifications. But as a representative battle for the early war period it could not easily be bettered.

In the course of my (admittedly somewhat random[32]) research, however, I have come across forty-one other references to the range at which fire was delivered in the battles of 1861–2 – the early war period. Of these 41 only five give both a 'maximum' and a 'decisive' range (400 and 25, 100 and 10, 150 and 25, 200 and 50, 400 and 75 yards respectively), but the general average works out at 122 yards, or almost twice the Seven Pines result. This includes only eight references to firing over 200 yards, and eight more between 101 and 200. Well over half of the references are therefore compatible with 'Napoleonic' ranges of a hundred yards or less. If the seventeen references from Seven Pines are combined with the forty-one other 'early war' results, we find the average range itself is only 104 yards.

If we look to 1863 as a putative transition point in the war – the mid-war period – we find that there were far fewer battles and therefore fewer references to range data. I happen to have encountered only ten, including two with both 'maximum' and 'decisive' figures (200 and 80, 40 and 20 yards). The average works out at 127 yards, with five references – exactly half – mentioning a range greater than 100 yards. The longest range mentioned is 250 yards, however, which contrasts with a reference to 500 yards in the early war sample.

In the late war period of 1864–5 I have encountered forty-five references to range, including six with both end points indicated (400 yards and 'close', 300 and 50 yards, 300 yards and 'close', 40 and 10, 120 and 10, and 200 and 30 yards. I have taken 'close' to represent 20 yards in these cases). The average works out as a range of 141 yards, or rather more than for the two earlier periods, albeit still below the 200 yards quoted for the Brown Bess. In fact there are still only nineteen references to fire over 100 yards range – less than half the sample – with the longest ranges mentioned being 500 yards once and 400 yards four times. These statistics are summarised in Table 6.1. The general picture that they paint is that ranges did gradually lengthen as the war went on, but that they never lengthened very far. The difference between 68 yards for a smoothbore musket in a wood in 1862 and 141 yards for an Enfield fired across a clear glacis from an earthwork in 1864 is not particularly impressive, and the general difference between 104 yards for 'early war' and 141 yards for 'late war' is less impressive still. It is particularly striking that out of a sample of 113 references I have encountered only 17 which mention ranges longer than 250 yards, and none beyond 500. This contrasts with a well documented army corps volley at 1,000 yards at the Battle of Talavera in 1809, at a time when Monsieur C. E. Minié was still only five years old.

It is difficult to find any evidence at all to support the suggestion that Civil War musketry was delivered at ranges much longer than those of Napoleonic times, at least if one excludes the very tiny number of snipers who did occasionally fire beyond 500 yards. When it comes to the fire of

TABLE 6.1 RANGES OF MUSKETRY FIRE

	Seven Pines	Other 1861–2	These two combined	1863	1864–5	Total
Total references in sample	17	41	58	10	45	113
Total ranges mentioned, in yards	1,033	5,002	6,035	1,270	6,340	14,645
Average range in yards	68	122	104	127	141	127

formed bodies acting in concert, we have to admit that the arrival of the rifle musket actually made very little practical difference – whatever may have been its theoretical potential to revolutionise the battlefield.

It is possible to argue that the limited but nevertheless remorseless increase in battlefield ranges during the Civil War was the result of gradually improving weaponry, but it was more probably the spread of disillusionment later in the war that was really to blame. In every era of military history long-range fire is the preferred tactic of soldiers who have lost their taste for assaults at close quarters – and the example from Talavera, already cited, was very much a case in point.

We should also remember that there is a definite limit to the usefulness of long-range fire even today. Even with fully motivated troops the opportunities for fire over two or three hundred yards are rare in modern battle. Studies of the Second World War and the Korean War suggest that normal combat ranges were about 100 yards, and in Vietnam the figure was still lower.[33] There is therefore a fallacy in the notion that longer-range weapons automatically produce longer-range fire. The range of firing has much more to do with the range of visibility, the intentions of the firer and the general climate of morale in the army.

In the Civil War there was certainly a recognition that fire at excessive range would be wasted. In 1864 21st Virginia refused to imitate its neighbouring regiment, which fired at 400 yards, for this reason.[34] 35th Massachusetts in 1862, despite being armed with Enfields, considered that 300 yards was out of range of the enemy. Its main line lay down in the open at this distance, sending skirmishers nearer.[35]

It seems that it was only by the time of the Franco-Prussian War of 1870–1 that genuinely long-range musketry was sometimes deliberately used. At Gravelotte and Mars la Tour we hear of massed volley fire from Chassepot rifles falling around Prussian units at ranges of more than a kilometre. Although this fire was unaimed and rarely deadly, it created an effect and could even sometimes turn back an infantry column or a gun battery.[36] There was nothing like this in the Civil War, even though the Springfield and Enfield rifle muskets might theoretically have been capable of something approximating to it. Such long-range massed fire was always very wasteful, and depended for its effect as much upon high volumes and rapid reloading as upon long range. The Civil War rifle musket could not be fired at anything like the same rate as the Chassepot, and commanders therefore preferred to employ it at ranges at which they had a reasonable hope of scoring hits. Regardless of whether they favoured the 'firepower' or the 'shock' theories of tactics, that usually meant something less than 150 yards, and sometimes as little as six feet.[37]

Some modern commentators have complained that this failure to exploit available technology was foolhardy or inexcusable, and many more have condemned the use of heavy 'Napoleonic' formations so close to the enemy. Yet if the standards of marksmanship and the general efficacy of fire had really not much improved since Napoleonic times one can scarcely quarrel with the continuing use of Napoleonic concepts. Only hand-picked and expertly trained shots could have hoped to score hits on the Civil War battlefield at much more than 150 yards, but such soldiers were as hard to find as the ammunition needed to give them target practice.

It may be true that advancing infantry perceived themselves to be at risk from musketry at longer ranges in the 1860s than they had done fifty years earlier, since undisciplined defenders opening fire at excessively long ranges may have caused bullets to fall harmlessly nearby. In particular, the peculiar whistling sound of the Minié bullet[38] must have made the fact of incoming fire more noticeable. Hence the 'zone of fear' for an attacker was perhaps somewhat extended beyond Napoleonic norms. Nevertheless, it is clear from the record that the real effectiveness of musketry was negligible at all but point-blank range. Even at the noted 'slaughter pens' at Bloody Lane, Marye's Heights, Kenesaw, Spotsylvania and Cold Harbor an attacking unit could not only come very close to the defending line, but it could also stay there for hours – and indeed days – at a time. Civil War musketry did not therefore possess the power to kill large numbers of men, even in very dense formations, at long range. At short range it could and did kill large numbers, but not very quickly.

Where we find a reference to both the 'maximum' and the 'decisive' range of fire, furthermore, we usually find that the latter is very short indeed. Out of my fourteen examples the average is 33 yards. This suggests that the existence of a wide field of fire may be regarded as irrelevant to the effect of musketry in the Civil War. All that was needed was a clear shot over perhaps the last 75 yards, as Captain Nisbet had arranged for his men at Resaca, or the last 55 yards, as General Griffin had contrived in the Wilderness.[39] This was the concept of tactics understood by Civil War officers who bothered to think about the matter – not some fancy idea of using the rifle musket's power at long range.

In order to determine the effect of open fields of fire upon the range of musketry it may be worth considering the case of Antietam, one of the most open battlefields of the war. I have found eleven references to range in this battle, giving an average of 107 yards. Only four ranges are longer than 100 yards, however, with the longest standing at 300 yards. Admittedly this was an 'early war' battle in which considerable numbers of smoothbores were used, but even so the ranges are not as long as the terrain might lead one to expect. The contrast with the figures for the more wooded combat at Seven Pines is not dramatic.

It is rather harder to find such an open battlefield in the late war period, since trees and fortifications were then the normal rule. If we take the reasonably open terrain at Cold Harbor and Petersburg, however, we find ten references to range, with an average of 186 yards, and only two references to ranges below 100 yards; 400 yards is the longest, and it is mentioned twice. This does admittedly represent a considerable extension of the field of fire, although the average still falls short of the 200 yards claimed for the Brown Bess.

It was apparently even harder to find open fields of fire in the Western theatre than in the Eastern, since the overall average range in the West, from my sample, was exactly 100 yards. That in the East was 136 yards. However, my sample contains only 22 references to ranges in the West as opposed to 87 in the East (and 4 unclassifiable). This doubtless betrays a certain weighting of my sources in favour of the Virginia theatre, although it is also quite probable that fewer firefights actually did take place in the West – and certainly that any given individual was far less likely to see as many there as in the East.

VARIETIES OF ATTACK

Now that we have established that Civil War earthworks were seldom barriers to movement, and that musketry was deadly only if maintained for a long period of time at short range, we can see that the success of the defence in this war was based largely on bluff. If two regiments opposing each other could often fire off their entire load of ammunition without reaching a conclusion, it is unlikely that a regiment holding a line could literally shoot down a massed attack. In the time the attacker needed to cross the last 33 yards – a minute at the outside – a defender who had already fired a few shots would be incapable of bringing down enough accurate fire to hit more than a small proportion of the assailants. The attacker would surely lose far fewer men if he pushed on without stopping than if he halted and tried to reach a conclusion by protracted volleying. If more Civil War regiments had understood this principle and possessed the military qualities needed to carry it out, we would doubtless have seen far fewer indecisive outcomes to the firefights.

Nevertheless, the bluff was usually effective and, for a variety of reasons, the attacker often did stop short of his objective. We even occasionally encounter the paradoxical spectacle of a regiment holding a position for some hours after it had used its ammunition.[40] Certainly a defender often fired high, not daring to lift his eyes above the protection of his works[41] but depending for his effect upon the flash and crash of the detonation, and the enveloping smoke which mercifully hid the scene from view.

On the attacking side there were many pressures acting against a resolute assault, apart from the enemy's appearance. The gruesome spectacle of the attackers' own wounded and dead would naturally create a deep impression, while the disintegration of drill would remove an

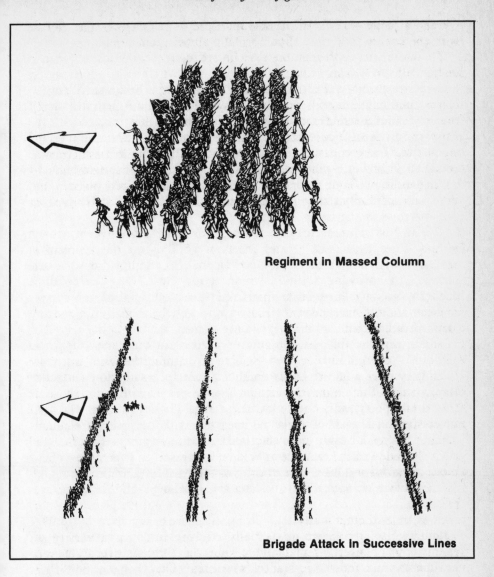

Regiment in Massed Column

Brigade Attack in Successive Lines

important motor of unified action. The insidious doctrine that battles were supposed to be fought by firing must also have created a psychological barrier to forward movement – and in any case it was always easier to stand still doing something familiar than to venture into the unknown.

Some tacticians tried to overcome the natural inertia of units in combat by the use of shock tactics, as we have seen. Others, however, sought their salvation in inventive assault formations. Taking their cue from Napoleon, Jomini and the whole geometry of engineering science, commanders

151

devised a range of different layouts intended to bring troops face to face with the enemy in a more than normally advantageous manner.

The basic infantry formation was a line of regiments deployed two men deep, with skirmishers ahead to prepare the way and supports a few hundred yards in the rear to cover for accidents.[42] In the classic European conception these intervals would be observed throughout the battle, and if the first lines needed succour they would be briskly replaced by the supports. In the difficult terrain of many Civil War battles, however, the succeeding lines tended to bunch too closely together until all became confused in a single shapeless mass. Attempts to meet the problem by adding additional support lines in the rear only made matters worse, as did the practice of closing the theoretical intervals between lines to 75 yards or so.

The use of columns instead of lines was more attractive in broken terrain, since they gave greater manoeuvrability, but the problem of excessive bunching remained. Once again, the attempts to solve the problem by throwing additional men at it – by making the columns deeper – was a failure. In any columnar assault formation the correct solution should have been to find ways to get the front ranks moving forward rather than to build up a mass of men behind them.

Some large assaults in columns or a succession of lines met with success[43] – enough to keep the idea alive throughout the war – but more often they were a failure. Most notably successful was Upton's attack at Rapahannock Station on 7 November 1863, repeated with much the same troops at Spotsylvania the following spring. The layout here was four successive lines each of three regiments – with the leading elements detailed to break outwards to the flanks once the enemy's first line had been cleared, and the rear line to hold back in reserve. This represented a move towards specialised assault drills of some sophistication, although it did not always work. A few weeks later at Cold Harbor Upton's attack was repulsed.

Sometimes there was an attempt to strengthen a heavy, narrow-frontage assault by placing columns on the flanks ready to turn outwards to beat off enemy counter-attacks – a technique somewhat dubiously recommended by Jomini and endorsed in Halleck's writing.[44] On other occasions there were efforts to imitate Napoleon's favourite 'mixed order' by placing some units in line and others in column. In general, however, the lesson learned by experience was precisely the same as had been learned in Napoleon's battles fifty years before – namely, that any form of 'monstrous column', whether it was composed of regiments in column or line formation, was exceptionally vulnerable and unwieldy when close to the enemy. This was no new perception born of improvements in small arms; it had been the most fundamental teaching of the European theorists since 1815. The American generals who saw fit to ignore it could doubtless be accused of following outdated foreign practice, but it was abusive practice which had long been superseded in the more advanced schools.

French theorists such as Marmont, Jomini, Morand and Pelet looked not to heavier formations but to lighter ones. They wanted attacks to be made in a line of small columns – each no more than a regiment – with wide deployment distances between them and supports kept well to the rear until needed. This was a tactical conception which placed the responsibility for victory upon the individual regiment, and even the individual company. It did not submerge them in an anonymous mass of from ten to fifty thousand men thereby relieving them of any sense of personal trust or accountability for the result. To judge by the Civil War memoirs, troops most appreciated it when their regiment was left to fight its own battle in its own way, receiving support only when all else had failed. Many of the most brilliant actions of the war were conceived and executed on a small scale, drawing the best out of every individual soldier.

Doubtless Burnside at Fredericksburg or Pickett at Gettysburg would have replied that a large-scale enemy position could not be carried by a series of disjointed small-scale actions, even if each one of them could be guaranteed success. There were too many opportunities for local counter-attacks by the enemy to disrupt progress and reverse the advances piecemeal. Only a massive single move against the key point could be made to stick within the time-scale available. Any other policy would have initiated an uncontrollable series of scrappy little actions which might have continued for a week.

The answer to this superficially attractive argument is that by 1864 the American battles actually had started to have the character of a long series of small disjointed attacks, although by that time there was admittedly also a breakdown of unanimity and determination. Hence the best results could not be achieved by this approach, as they might have been if it had been applied earlier in the war. But at least by 1864 the higher commanders had learned to call off mass attacks which went wrong. Sherman at Kenesaw and Grant at Cold Harbor both called an end to the fighting considerably earlier than had Burnside at Fredericksburg.[45]

The trouble with these mass assaults was that they became disorganised by their own inner mechanisms, almost independently of enemy action. Hence the supposed advantage of superior numbers actually became a disadvantage, because no one was left free to fight in a scientific or carefully controlled manner. We therefore see the paradox that a smaller attack against a strong enemy position could be more effective than a large one. Once again, the best hope of safety lay in the apparently more dangerous course. If the time required turned out to be longer than the few minutes notionally needed for a 'monstrous' column, then that was a cheaper price to pay than the failure and ruin which usually attended the latter.

It is perfectly fair to point out that these lessons had also been forgotten by the European generals of the First World War, and that their massed set-piece attacks in the early years were only too slowly superseded by more flexible small-unit tactics, infiltrations, and so forth. Yet there was also perhaps a difference between the two cases, in that the battlefields of the Western Front did display some genuinely new features which took time to understand, and which had not been adequately thought through in advance. In the case of the Civil War, by contrast, there was little that was new – but a large body of relevant pre-existing theoretical analysis that was misunderstood or badly applied. In the event Halleck's warning of 1846 was confirmed:[46] improvised American militia armies did indeed fail to master the art of mobile tactics – though he was probably wrong to think that they were inevitably incapable of doing so.

If 'monstrous' columns represented the tactics of the past and light columns represented those of the present, it was still lighter and more dispersed formations that represented the tactics of the future. The idea of an attack entirely in skirmish order had been well known for at least a century before the Civil War, and frequently applied in Napoleonic times. Much of the discussion in the *chasseurs à pied* had concentrated upon it, and hence it was incorporated into Hardee as one possibility open to the Civil War commander. What was rather new in the 1860s, however, was the idea of basing an entire battle upon it, even in open ground without solid supports close at hand. This went beyond Hardee and reflected some of the more advanced thinking of the *chasseurs*.

At the beginning of the Civil War some regiments had adopted the so-called 'Zouave rush' as their tactic.[47] Instead of advancing in a close line, stopping to fire from time to time, they charged forward in loose order at the quickstep and threw themselves flat between bounds, both to fire and to regain breath. Whenever they presented themselves as a target to the enemy, it was a difficult one to hit. This was an exciting and sensible way to fight, although it did perhaps suffer from a certain lack of ensemble for the final bound into the enemy's line. It was used with success by such diverse military organisations as the Garibaldini when they landed in Sicily in 1860, and by the Prussian Guard following their failure at St Privat in 1870.[48] In America it was identified with the way the Indians had fought, and was soon terminologically appropriated by New World nationalists. Instead of the 'Zouave' rush it became the 'Indian' rush.

Some authorities have claimed that the Indian rush eventually became the predominant assault tactic of the Civil War,[49] although the evidence for this appears slight. It may well be that 'formless and indecisive skirmishing' was the predominant method of action in the war, but that is not quite the same thing.

There is admittedly a problem of definition in all this. To many observers almost any infantry which lost its drill-book precision, moved independently and used cover – with each man 'yelling on his own hook and aligning on himself' – seemed to be behaving in an 'Indian' manner. Already at Gaines's Mill one Union survivor was able to report that 'The firing of the Texans was so accurate and their movements so cunning and Indian like that [I] never wish to make their acquaintance again'.[50] In this battle the initial Texan attack had a certain amount of cover to exploit, but much of the way was across an open field, which surely precluded 'cunning' movements. Nor was Hood anxious to see his men stop to fire, let alone indulge in stealthy stalking and sniping. We must conclude that this 'Indian' behaviour of the Texans must have occurred somewhat later than their initial assault in close order, and was more in the nature of conventional skirmishing than a true Indian rush.

In 1864 on the Opequon, Captain De Forest had taken part in a static firefight in which he noted the deliberate and accurate fire of his opponents at a range of 200 yards. His conclusion was that the Army of Northern Virginia

... aimed better than our men; they covered themselves (in case of need) more carefully and effectively; they could move in a swarm, without much care for alignment or touch of elbows. In short, they fought more like redskins, or like hunters, than we.[51]

Once again, this perception seems to be based more upon Confederate skill in firing from cover than upon their offensive tactics, and the idea of a 'swarm' suggests a skirmish movement more than a full-blown assault. At Winchester in 1864 'One of Jackson's Foot Cavalry' observed such an action from afar and applauded each bound that brought the skirmishers to a new piece of cover nearer to the Union line.[52] But it still did not produce a final charge to overrun the position.

Skirmishers *could* sometimes make a charge, as 22nd Massachusetts and 4th Michigan had done at Spotsylvania.[53] This was an attack without fire and with bayonets fixed, and it successfully cleared the first enemy line. An attempt to go further against a stronger body was nevertheless turned back with heavy loss. There is no mention of moving in short bounds in this description, however, so it still does not quite fit into the strict definition of an Indian rush.

Most Civil War battlefields were pulsating with skirmishers, especially by 1864. Anything between one and all ten companies of a regiment might be detailed for the duty, and often there might be two regiments sent forward from each brigade. Yet the standard procedure for these skirmishers was to do no more than probe the enemy, occupy his attention and maybe cause him some loss. At Fredericksburg 13th Massachusetts enjoyed especially fine drill training in skirmish movements, and prided itself upon its efficiency in this type of fighting. It advanced, drove off the enemy skirmishers by fire and then used all its ammunition against the enemy's main line – but it did not assault. It left that task to the heavy

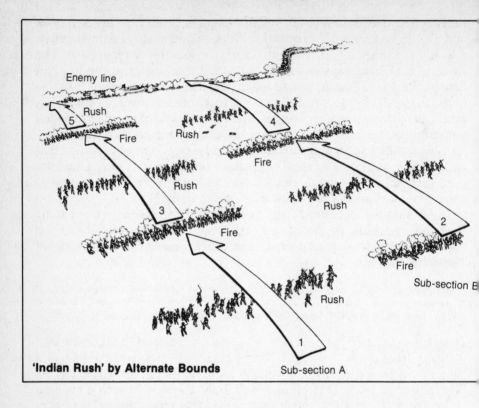

'Indian Rush' by Alternate Bounds

supports moving up from the rear, and gratefully retired once they had passed through.[54]

By 1864 we often find the same pattern repeated, but more often than not without the final appearance of heavy supports. Typically the skirmishers would move forward, probe the enemy entrenchments and find them too strong. They would then invoke this as a reason for cancelling the main attack before it even began. Sherman called such action 'Indian war', but it is clear that it rarely contained any serious aspirations towards a decisive result.[55]

Without the dynamic of a genuine attack across the final 33 yards, we cannot really call skirmish action an assault technique. We are therefore left with relatively few examples of a true skirmisher attack, and still fewer of a thoroughbred Indian rush. Apparently one of the latter was performed by Sigel's men on the second day at Pea Ridge in Arkansas – a battle at which there were even some real Indians present. At Belmont Grant's men also moved to the attack from tree to tree.[56] But most of the other claimed instances of this technique ultimately appear to have been instances of something different after all.

At Beaver Dam Creek in 1862 there was at least one attack by a formed

156

body which threw itself to the ground when the enemy fired his salvo, and then jumped up again to carry the position. But this was against a single artillery piece, where obviously the timing of each shot could be exactly predicted.[57] Against an infantry regiment giving a continuous fire at will such virtuosity would have been impossible.

One undoubtedly very significant Civil War assault technique which we have not yet considered is the trench raid. Take for example Captain Nisbet's exploit on the Kenesaw line in 1864, which stands in such stark contrast to the disastrous massed Union attack near the same spot a few days earlier. One night the Confederate picket line was manned with specially selected volunteers:

Armed with Colt's navy seven-shooters, self-cocking 44-calibre, they had instructions to crawl from our works and drop into the enemy's vidette pits, which were not more than fifty feet from our videttes. Each man was instructed to present his pistol to the vidette's head, and tell him (softly) he would blow his head off if he made any noise *and to do it* if need be.
The night was rainy and very dark. The scheme was successfully executed . . .[58]

After this some 200 shock troops rushed through to the now-unguarded front of the Union line, and captured the best part of a regiment of regulars before the alarm had been sounded or supports sent forward. The Confederates had their prisoners safely back in their own trenches before the counter-bombardment began.

Nisbet's message is that much could be achieved bloodlessly by stealth which would otherwise require great sacrifice and waste.[59] 'Johnny Green of the Orphan Brigade' made a similar point when considering the heavy fog which often engulfed Missionary Ridge at dawn:

. . . it occurred so often that I wondered that the yankees did not come at us along with the fog, for in this way they could get right upon us before we would know it unless the noise of their coming should betray them.[60]

By chance this was precisely what Hancock achieved in his dawn attack on the Confederate lines at Spotsylvania, although it appears to have been neglected on many other occasions when it might have been decisive.[61] At Fredericksburg in particular there were recriminations among the Union troops that the thick fogs were exploited for retreat but not for assault.[62] In March 1918 this was not destined to be a mistake which the German shock troops would repeat.

Even without fog a dawn attack could win surprise and achieve impressive results. 123rd New York conducted a model trench raid at Atlanta, capturing 200 prisoners for no loss.[63] Equally, the indefatigable General Griffin at Petersburg managed to infiltrate an entire brigade by dawn to a hundred yards of the enemy line. The attack was conducted without firing, received only one artillery round from the Confederates and captured 600 prisoners and a battery.[64]

Night and fog ought certainly to have been exploited more often by Civil War tacticians, as they have been in the twentieth century even to the extent of creating artificial fog in the shape of smoke screens. Night and fog are none the less notoriously tricky to manipulate, since it is a part of their nature to blind and befuddle friend as well as foe. In the Napoleonic period there had been many an enterprising commander who had hoped to win a trick by a night attack, only to find that it backfired upon him. Wellesley at Seringapatam and Abercrombie in North Holland had both experienced this, and each had sworn never to take such a risk again. Then in the Civil War there were repeated attempts to march at night which also caused disproportionately high resentment and confusion in return for disproportionately low mileage completed. When it came to actually fighting at night, furthermore, it was usually at the end of a long day of combat, when the opposed forces were already exhausted and ready for an excuse to cease fire. Bartlett Yancey Malone of 6th North Carolina made a dusk attack at Gettysburg – 'But it was soon so dark and so much smoke that we couldn't see what we was a doing'.[65] This was usually the way with such ventures.[66] Friendly units stood more than a normal chance of firing into other friendly units, and even if initial success was gained it was exceptionally difficult to consolidate – or to keep hold of one's prisoners – in the gloom. When Jackson wanted a night counter-attack at Fredericksburg it was forbidden by Lee as too risky, and at least one of the attendant staff officers concurred with heartfelt relief.[67] Ironically, it was to be in just such an operation a few months later that Jackson himself was accidentally shot by his own troops.

Woodland fighting was already very difficult to control without the added hazard of low visibility to increase the risk of confusion or panic. Nevertheless, steady troops could sometimes win significant advantages by fighting at night or in fog. One case of this was the combat of Lookout Valley near Chattanooga, fought at the dead of night. At first the Union troops succumbed to the normal sources of confusion, including stampeding mules and Confederates pretending to be Federals; but they were later successful when they renewed their assault, this time without firing until after the enemy had run off. Then a few weeks subsequently Lookout Mountain itself was stormed in force, in daytime but in thick clouds. The attack was more successful than had been hoped.[68]

ACCELERATING AN ATTACK

Loss of impetus and failure to achieve shock were the main enemies of the Civil War tactician who wanted to cross the vital last 33 yards to come to grips with his foe. Instructions not to fire – or sometimes not even to load – were a common response to this need, although they were obeyed less often than the officers might have wished. The use of massed formations turned out to be even less successful, while raiding covered by night or fog was a specialised operation which could often go wrong. In many firefights commanders had to rely instead upon rather simpler expedients, which, although they were individually quite trivial, might in combination add up to something imposing enough to chase the enemy away.

The first consideration was the simple one that troops committed to a charge should be fresh and well in hand, preferably having eaten not long before. It was failure to observe this principle that had brought disaster to the British at New Orleans, as it was to bring equal disaster to many a Civil War regiment thrown carelessly into a fight in unpromising conditions. For example at Culp's Hill, Gettysburg, Steuart's brigade had been in action for six hours, and had run out of ammunition at least once, when it was asked to make a very difficult uphill attack against superior numbers in strong fortifications at right angles to its own line. The attack was made without any support from infantry or artillery, but under enfilading cannon fire.[69] If it failed we can scarcely attribute that to 'newly improved musketry': it was due merely to neglect of the most basic rules of warfare. That some members of this attack came to twenty yards of the enemy line shows what might have been achieved by a better-planned operation.

Another obvious course of action was for an officer to make the maximum use of his vocal chords:

In battle the order to charge is not given in the placid tones of a Sunday-school teacher, but with vigorous English, well seasoned with oaths, and a request, frequently repeated, to give them that particular province of his Satanic Majesty most dreaded by persons fond of a cold climate.[70]

Modern studies of combat psychology have shown that constant chatter in a fighting unit helps to dispel the loneliness of fear, and to strengthen cohesion. Civil War officers who habitually shouted themselves hoarse in battle were therefore almost certainly doing the right thing, even though the deafening roar of combat might drown them out – and on at least one occasion a regiment advancing through a cotton field made earplugs for themselves to deaden the din.[71]

Nor was it only the officers who were supposed to shout, since whole regiments were encouraged to do so:

At the same time you are ordered to yell with all the power of your lungs. It is possible that this idea may be of great advantage in forcing some of the heroic blood of the body into the lower extremities. Whatever may be the reason, it was certainly a very effective means of drowning the disagreeable yell of the enemy.[72]

The above may be termed the 'defensive' use of yelling, to help one's regiment remember its identity. The 'offensive' use could be still more effective, as a weapon to chase off the enemy. We read that in the Wilderness battle:

... the yellers could not be seen, and a company could make itself sound like a regiment if it shouted loud enough. Men spoke later of various units on both sides being 'yelled' out of their positions.[73]

At Malvern Hill the Confederate orders were to 'Charge with a yell'[74] if the enemy started to break, and this was fairly standard practice in the assault. A yell created a fearsome impression just as much as a volley held to close range or a silent onset of determined men. Whether it was a demure New England 'cheer', a 'Hoosier yell', an 'Indian war whoop', a 'Texas war whoop', a 'regular Mississippi yell', a 'Go in Baltimore!', or 'three times three for Tennessee'[75] did not particularly matter: the effect was the same. Wellington's men had used carefully timed cheering as a part of their shock tactics fifty years earlier, and it became yet another part of Napoleonic warfare which was carried over into the Civil War without diminution.[76]

Another way to speed an attack was simply to lighten the men's load. The soldier on campaign tried to carry as little impedimenta as possible, but apart from his weapons there always remained an irreducible minimum of blankets, food, spare clothing and keepsakes. In each messing group there were communal cooking utensils which were particularly large and unwieldy. In the attack on Marye's Heights at Fredericksburg a company commander in 22nd Massachusetts reports:

I turned to see if my company kept its formation intact, when Ned Flood, custodian of the treasured frying pan, held it out so as to catch my attention and asked by word and gesture, permission to drop it. I assented. With a serio-comical look of grief he cast it from him ...[77]

In the same advance a sister regiment, 35th Massachusetts, had earlier felt the effects of its impedimenta even more:

The regiment ... passed on, getting breathless with the run and their burdens, men dropping by the enemy's fire all along. We reached a wire fence ...[which] broke the formation of the line, the wires catching a man by some of his many bags and bundles, and persisting in holding him until he unslung the impedimenta, or was extricated ...[78]

Later in the day the 35th was to be relieved by the 22nd in 'admirable' manner, and retired at dusk across the ground won. There the men made good their material losses from the 'equipments and equipages of all sorts'[79] which littered the battlefield. Including, presumably, Ned Flood's frying pan.

Our two Massachusetts regiments at Fredericksburg appear to have gone into action with all their campaigning gear draped around their persons; it was hardly surprising if they found it heavy and cumbersome. On many other occasions, however, Civil War units were ordered *en masse* to deposit their baggage either before entering battle or, more heroically, before launching into some especially daring venture once battle had been joined. 'Packs down – charge!' was almost as stirring a cry to the weary warrior as the more conventional 'Fix bayonets – charge!', and certainly signalled an even more serious intention. To throw down one's pack, especially if no arrangements were made for guarding or recovering it, was equivalent to instant destitution for the soldier. Only the most complete victory could make good the loss, so the troops could be expected to fight with the reckless abandon of those who have nothing further to lose.

The timing of a 'packs down' order was a matter of delicate timing since if it was given prematurely, before the need arose, it would cause bitter resentment among the troops.[80] Even the best arrangements for retrieving personal baggage which had been dumped were notoriously unreliable, as 35th Massachusetts discovered when they had 'their' packs forwarded from the Arlington training camp to the front at Antietam.[81] Conversely, a refusal to allow troops to jettison their equipment before a strenuous combat could cause equal resentment, not to mention some heavily overloaded soldiers whose combat value was thereby reduced.[82] Not only would they be more ready to run away than they might otherwise have been, but their first act upon doing so would be to jettison their hated baggage.

The battlefield would thus be doubly littered – partly by the soldiers moving into the fray, and partly from those running out of it. The man who could hold onto his pack throughout a battle could consider himself quite fortunate, although for the rest there would often be plenty of items available for salvage. If only one could reach the prisoners and casualties before the professional scavengers and 'bummers' had a chance to ply their ghoulish trade, one could fit oneself out as good as new.

Before we leave this subject it may be as well to remember that the promise of plunder could itself be a spur to assaulting troops. The deprivations of campaigning life were considerable in this war, in which many soldiers went without a proper roof over their heads for their full three years' service, and whole armies sometimes starved. There was more than a grain of truth contained in General D. H. Hill's anecdote of the wounded Irish soldier at Chickamauga who urged on his companions by shouting 'Charge them, boys! They have cha-ase [cheese] in their haversacks!'[83]

In Napoleon's day the motive of plunder had been no less great than in the Civil War, but the packs-off charge had been less common. At least a part of the reason for this must be found in the generally slower pace of battlefield manoeuvres, and hence the relatively lesser exertions demanded of an attacker. By the 1840s, however, the rise of *chasseur à pied* theories of the *pas gymnastique* had started to change this picture, and in the European wars of the 1850s it seems that the packs-off attack was already quite a normal feature. When we come to the 1860s we certainly find that the American armies often used Hardee's double quick time or even the dead run in the attack, and hence must have been more than ready to unburden themselves of their packs. They were attempting to manoeuvre around the battlefield faster than Napoleon's soldiers, faster than Scott's in Mexico and faster than Halleck had assumed was even possible for fragile militia troops.

After the war Upton retained the double quick time of 165-180 paces per minute in his tactical manual[84] so he, at least, must have felt it was worth while. He also retained the idea of jogging over much longer distances than a charge (4,000 paces in 25 minutes). Nevertheless, he appears to have been ambivalent about the tactical value of such high speeds, and like Ardant du Picq in France he apparently believed that a running unit would be especially fragile. He advised skirmishers that 'the very ardor with which an enemy pursues a temporary advantage will surely secure his defeat, if boldly and unexpectedly confronted by the men whom he had supposed to be demoralised'.[85] In a battalion attack Upton recommends a combination of gaits so that the colonel can maintain his control up to the final rush. The battalion should try to cross the beaten ground as fast as possible, but should not break into the double until all the men have a chance to arrive on the objective together.[86]

Such caution on the part of officers seems to have been commonplace in the Civil War itself. It was less difficult to persuade troops to run forward in the attack, apparently, than it was to keep them together under unified control. Captain De Forest's regiment, which had been so proud of its drill in its first battle, could maintain no more than 'a loose swarm' in 1864 as it doubled a quarter of a mile forward over an open field to take up a firefight from a demoralised line. During this advance each individual apparently stopped occasionally for breath, and some attempted to stay under cover rather than to continue forwards.[87]

At Gettysburg Hood's troops found that their attempt to double forward 400 yards on a hot day left them exhausted for the fighting at the far end. At Seven Pines 17th Virginia doubled a mile and a half to the battlefield, regrouped and drew breath behind a woodpile, then made a charge. Whether it was because the men were blown or because fugitives retired through the ranks, the compactness of the regiment was lost. A few weeks later at Frayser's Farm the same regiment had been warned not to double forward until the enemy was in sight, but it did so anyway while there was still almost a mile yet to cross.[88] Presumably it was something similar to this which led Sergeant Hamlin Coe to make the startling comment about his battle at Adairsville in 1864 that 'Although our regiment was broken and in disorder, they charged like tigers'.[89]

It is at this point that the disciplined jogging of Hardee's double quick time must have merged into the formless mob tactics of a skirmish attack or even the elusive Indian rush itself. It is more than probable that many attacks which started off as orderly drill formations, well under the hands of their officers, degenerated into a swarm of individuals moving forward with enthusiasm but not cohesion. If these individuals eventually consolidated themselves into a thick skirmish line firing at the enemy from a distance, rather than sweeping over him in a unified movement, that was scarcely a surprising outcome. It showed that Hardee (and behind Hardee the whole *chasseur* movement) had been mastered only in part. It showed that a packs-down cheering charge at the *pas gymnastique* might work for Indians, Celts, or even (Halleck notwithstanding) for American militia in their first flush of enthusiasm, but that such tactics needed to be based upon a lot more specialised training if they were to work against formidable opposition.

This was never a problem that was properly identified in the course of the Civil War itself, nor was it very well identified elsewhere.[90] It was a major failing of the French *chasseur* school that for ideological reasons it gave inadequate attention to the weaknesses in its own formula, just as it was the fault of its opponents that they concentrated only on the style rather than the essence of what the *chasseurs* had to say. *Chasseur* tactics were actually enormously demanding, in terms of the assumptions of most nineteenth-century warfare, and they could not be transplanted easily or quickly upon an army which lacked either appropriate indoctrination or ammunition for target practice. Obviously one could not train men to shoot unless they were given hundreds of rounds to practise with, and one could not train them for the formal packs-down charge unless they went through the motions of it many times over in the roughest country available. None of this was practical in the Civil War, so the armies either improvised ineffectively or reverted to outdated tactics disastrously. They were not usually able to use the most modern tactics available to them – which ironically were incorporated in their own drill manuals – nor did they often hit upon the still more futuristic techniques of the trench raid or the infiltration.

7

Artillery

Civil War artillery has acquired a bad name for very much the same reasons as the bayonet. It is thought to have caused few casualties or decisive results. It is believed to have been outdated by advances in small-arms technology, and it is seen as an alien element imposed artificially upon infantry regiments which had little desire for its assistance, and less need. It was a remote, theoretical part of the battle which had something to do with either 'the rear' or 'the distance', but not with the firefight of real soldiers.

The use of artillery did not, of course, have the same character as that of the bayonet, but we can positively state that both of them were rather more important to the Civil War than is generally acknowledged, and that both of them have been similarly shamefully treated in the literature.

THE LONG ARM OF HUNT AND THE SHORT ARM OF LEE

At the start of the war the artillery of both sides was split into self-contained batteries, and each battery allocated to a particular brigade, regiment or even battalion of infantry.[2] The local infantry commander was therefore the commanding officer of the artillery, outranking the battery commander and taking most of his important decisions for him. The result was that guns were not used in the most advantageous manner, and their officers were overlooked for promotion. Many budding former West Point gunners therefore transferred to the infantry in the hope of making a better career.[3] Some of them certainly succeeded spectacularly well and we find, for example, that the pre-war 3rd US artillery regiment had at one time or another included all of the following great names: Meade, Sherman, Thomas, Reynolds, Bragg, Early and Burnside. The great artillery virtuoso Henry Hunt was another of its products, as was the gun-founder R. P. Parrott;[4] but the disproportionate number of famous non-gunners from the regiment is eloquent of the problem which the artillery faced.

165

The point is encapsulated in the career of the Union captain Stephen H. Weed. He was an officer who fought his guns brilliantly in the first two years of the war, and at Chancellorsville even commanded the artillery of a whole army corps. Hunt singled him out as having a particular flair for handling large masses of cannon, and wanted to see him promoted. Others, however, also recognised Weed's talents and made sure that he was indeed promoted to brigadier – but in the infantry. His services were lost to the arm he knew best.[5]

The source of many problems was that for much of the war commanding officers persisted in regarding artillery as merely a subsidiary technical branch, an auxiliary which might add a little extra vitality to a firing line if conditions were favourable – but more typically would not. They lacked a conception of the artillery as a force to be reckoned with in its own right, a force that could sometimes win battles unaided, if only it were released to follow its own tactics in its own way.

This perception is vividly reflected in the table of appointments for the artillery of the Army of the Potomac at Gettysburg, which included approximately 8,000 men with 372 pieces[6] – almost the manpower (and certainly the firepower) of a complete army corps. It included only two general officers, who complained they were used 'as messenger boys' and overruled by any passing infantry commander; then there were three colonels and no other high ranks at all. One army corps had its guns commanded by a lieutenant.[7]

An organisation the size of a corps was thus given a staff normally allocated to a brigade, and this state of affairs scarcely changed throughout the war.[8] Only gradually and painfully, through a long process of bureaucratic in-fighting, did the artillery succeed in winning some measure of recognition for its independent status and tactics. After Gettysburg the army's artillery commander was accepted as having an overriding authority in gunnery matters, with the infantry relegated to merely a consultative role, although in practice the change brought little real improvement.

Correct artillery tactics demanded that several batteries should be concentrated together on key terrain features, to mass their fire against selected targets. They should not be distributed piecemeal along the infantry line, merely boosting the firepower of individual brigades and regiments, but should be carefully co-ordinated at points where the battle as a whole could best be influenced. This often meant that enfilading cross-fires could be arranged, giving much more destructive effect than direct frontal fire at the same time as they relieved the gunners from fears of being personally overrun. Such fire, however, meant that local infantry commanders had to agree to 'their' guns being used against an enemy who was not directly threatening their own frontage.

Massing and crossing fire was a lesson which had been learned in the French wars of the 1790s, and almost as slowly relearned in the Civil War.

Before the 1860s there had been very little practical American experience in handling large masses of guns, and if anything even theoretical opinion was set against it.[9] Unlike the development of infantry tactics, therefore, the progress made by artillery during the war represented a break from traditional habits of thought. Whereas the infantry generally proved incapable of applying the most modern theories, and reverted to earlier patterns of action, the artillery gradually moved closer to European thinking the longer the war went on.

Even at the best of times, however, the achievement was patchy. The leading modern student of Civil War artillery even reached the conclusion that 'the capabilities and limitations of field artillery in this war were outside the grasp of all but a few officers on both sides'.[10] Success therefore became to a considerable extent a matter of personalities. If a man like Henry Jackson Hunt was available to co-ordinate the guns, then all would go well.[11] If such men happened to be absent, however, great opportunities might be wasted.

There was also a problem of terrain,[12] since few battlefields provided clear stretches of open ground in the positions the gunners would ideally have wished. For every Malvern Hill there were at least two woodland fights with restricted fields of fire. In the Wilderness the Army of the Potomac left more than a hundred of its guns unused in both 1863 and 1864, although on both occasions a surprisingly high number of guns were also sometimes assembled in grand batteries. At Hazel Grove in the 1863 battle, and then near the Chancellor House itself, both sides managed to field concentrations of 36 pieces or more.[13]

Such concentrations could sway the course of battle quite dramatically, and the question of winning artillery superiority at a particular point was of much more than merely academic interest. If one could establish one's grand battery at a commanding spot before the enemy could establish his, one could in theory dismount his batteries piecemeal as they arrived and then wither any infantry trying to operate on that piece of ground. Winning the gun duel had to come first, but then the rest would follow automatically.

The Union armies usually possessed more rifled and heavy guns than their Confederate opponents, so in theory they could expect to win the counter-battery contest at long range, secure from retaliation. At Antietam and Fredericksburg they established very powerful batteries for this purpose, over a mile from the enemy's front line and in places as much as 3,000 yards.[14] This represented a distinct technological advance over earlier practice since, although such specialised weapons as the Villantroys mortar had ranged up to 6,000 yards, the bulk of Napoleon's field artillery had ranged little further than 1,000 yards. The advent of rifling thus effectively doubled or trebled the potential range of the field artillery.

At both Antietam and Fredericksburg, however, the long-range Union artillery achieved considerably less important results than had been hoped. Although it unquestionably established a superiority over the Confederates in the duel of gun against gun, this did not by any means win the battles. The Confederates drew back their cannon into folds in the ground at Antietam, and into entrenchments at Fredericksburg, and held their fire until the advancing Federal infantry came close. Then at the psychological moment the masked batteries were revealed. They set to work with canister and grape, adding additional firepower to the musketry of the defence. The overwatching Union batteries could do little to interfere, since in many cases their own infantry was too close to the target and obscured it from view.

At Gettysburg almost exactly the same thing happened in reverse on the third day of the battle. Alexander's massed Confederate artillery played upon the Union lines at long range in preparation for Pickett's great infantry attack. Because the defenders' batteries retired under cover, however, they were scarcely damaged. Then when the charging infantry came close the guns sprang to life and materially contributed to its defeat.[15] It was this which prompted Private Edgerton of 107th New York to write: 'There is one thing that our government does that suits me to a dot. That is, we fight mostly with artillery. The rebels fight mostly with infantry.'[16]

Probably the most effective way to use artillery in the Civil War was to keep one's guns under cover until the enemy presented a massed target of attacking infantry at medium to close range. Then, provided the ground was not too steep or too obstructed,[17] a great deal of damage could be done to him before he could get close enough to shoot back effectively.

In the days before indirect or blind firing was practicable,[18] the gunner's dream must have been to secure an uninterrupted shoot against an unmissable target from, say, 1,000 yards down to around 200 yards or the range of musketry. The hope would be to disorganise and repulse the attack within that zone by the fury of the cannonade, and in practice this was often precisely what did happen. Regardless of whether the artillery was supported by infantry or not, it frequently succeeded in chasing off infantry attacks upon its own position. The literature is too full of examples[19] for this point to be laboured, although it may be worth mentioning that such actions certainly give the lie to allegations that musketry could automatically clear the guns from the field.

Problems arose for the gunners, however, on those occasions when infantry succeeded in coming within *their* normal firing range, that is to say within about 200 yards. It was rare for such infantry to sweep over an active battery in their first charge, but if they went to ground and maintained skirmish fire for some time the artillery would start to feel the pressure. Isolated, unsupported batteries would be forced to retire hurt – and it is these cases that have created the impression that infantry had somehow won the upper hand over the guns. Batteries supported by other guns or friendly infantry would not usually be forced to retire, although they would bleed all the same. Normally they would continue to exchange blows, doggedly, for a considerable period until they were finally compelled to fall silent and pass the battle to their supports. In these circumstances they could at least be sure that their sacrifice was not in vain, and that they had made a major contribution to the success of their cause. If they were backed by an efficient higher commander, such as Hunt at Gettysburg, they would also know that a fresh battery would be fed into the line as soon as they retired from it.

In this type of fighting there was actually not a very great role for the famous new long-range rifled cannon. The effect of artillery came mostly at 1,000 yards or less – in other words very much the same ranges as had been familiar to Napoleon – so the Civil War commander was left with only a handful of specialist tasks for his longer-range pieces. They might, for example, be used for siege work or the defence of riverine fortifications against naval threats. They might be used to assassinate enemy commanders at shockingly long range – as they were against General Bishop Leonidas Polk near Kenesaw – or they could confer great advantages if a counter-battery duel was accepted by the enemy. But for stopping a massed attack the rifled cannon was much less useful than the smoothbore.

The rifle fired shells at such a high velocity that they often dug themselves into the ground and exploded at a harmless depth, if they exploded at all. Many shells were defective, especially in the Confederate service, and they usually had to be of an especially small calibre for use with rifled guns.[20] The small calibre also meant that these guns gave a poor performance with canister rounds, which left them with enviably high accuracy but lamentably low killing power – precisely the opposite qualities to those most needed to stop a mass assault.

The limitations of rifled artillery were fully understood at the time, and although the Confederates could never get enough for their needs the Union forces complained of an overabundance of this type of weapon. Early in the war McClellan had decided that the Army of the Potomac needed only one rifle for every two smoothbores, but he never succeeded in bringing the ratio down to less than two rifles to each smoothbore; nor could Hunt force it much lower when he tried to get rid of some of the rifles after the battle of Fredericksburg. At the start of the Wilderness battle smoothbores were still in a minority and it was only when Grant sent home 122 pieces, in order to disencumber his collapsing road network, that Hunt was able to seize a fleeting opportunity.[21] The achievement of parity between rifles and smoothbores came too late to have much practical effect, however, and the siege of Petersburg soon diverted the Army of the Potomac into unfamiliar types of artillery such as howitzers and mortars.

If the long-range rifle was the unwelcome *prima donna* of the artillery arm, the 12-pounder smoothbore Napoleon was the much-loved maid of all work – the close- and medium-range crusher which achieved the big results. This weapon was based upon a design recommended by Napoleon III, and was considerably lighter and more manoeuvrable than the 12-pounders which had been available to his more celebrated uncle. In theory it could range over a mile, although in practice it was relatively rarely used at much above 1,000 yards. As with the infantry's rifle musket, there was a very important difference between the range at which the manufacturer could hit a target in his proofing trials and the range at which the soldier in combat found it useful to try his skill. Also as with the rifle musket, we find a surprisingly large number of occasions on which these guns were fired in the very shortest zone of their range. Artillery fire was not unknown at 20 yards from the enemy, and it was common below 300 yards.

The main effect of artillery came at what may be described as 'canister range' – the last 300 yards to the gun, sometimes extending to 500 yards.[22] It was here that the flash and crash of the heavy Napoleons, firing two and a half pounds of powder with each detonation, could numb and stagger the enemy, even when they did not physically hurt him. Certainly the shock power of such fire was of an altogether higher order of magnitude than that of musketry.

Against all this we often find it alleged that musketry not only outranged the artillery, but also caused far more casualties. Some commentators even speak as if the physical effect of the guns was marginal.

Admittedly, Civil War artillery was ineffective when it happened to be deployed at the wrong time or place for a battle, or when its own tactical ideas were overruled by the infantry. But when it had a fair shot at the enemy, and if all else was equal, there is no evidence to suggest that it did not do at least as well as its Napoleonic predecessors. The Union medical statistics from the Wilderness and Cold Harbor operations of 1864 indicate that about 6 per cent of the casualties in that battle were caused by solid shot and shell from artillery,[23] to which may be added an unknown proportion – perhaps another 6 per cent – caused by musket balls fired from cannons in the form of canister. An estimated 12 per cent of casualties caused by artillery in the Wilderness is remarkably high for a battle in which neither side could often deploy its guns off the road, in which fields of fire were often no more than a few feet, and which one participant described as 'more exclusively a small arms fight than any battle of the war'.[24]

In the 1,000-yard 'shooting gallery' at Malvern Hill it is thought that 50 per cent of the Confederate casualties were attributable to artillery, and the same figure has also been claimed for Union losses at Fredericksburg.[25] Such figures show that the guns were disproportionately effective as killers – precisely as they had been in Napoleonic times. Even if we accept the lowest estimate of lethality – the 6 per cent proven casualties to artillery in the Wilderness – we still find that the guns almost managed to hold their own in the league table of destructiveness. The Army of Northern Virginia at this time had just over 9 per cent of its manpower in the artillery arm, so the casualties they inflicted were almost proportional to their numbers.

The effect of artillery in these battles should not, in any case, be measured in terms of casualties inflicted but – as with the bayonet – in terms of tactical advantages gained. Artillery had a shock power and a deterrent effect which is beyond the statistical reckoning. It could break enemy units and disperse attacks even when it did not hit enemy soldiers.

Colonel Wise, the Confederate historian of Lee's artillery, was perfectly correct to point out that

We often hear the sneering criticism that at such and such a battle but 1 or 2 per cent of the enemy's loss was due to the fire of the artillery. Any such test is entirely erroneous. Not only do the guns exert a tremendous moral effect in support of their infantry, and adverse to the enemy, but they do far more. They often actually preclude heavy damage from the enemy by preventing him from essaying an assault against the position the guns occupy. Then, again, by forcing the enemy to seek cover, they eliminate their antagonists to that extent... Let us hear no more of artillery efficiency as measured by the number of its victims.[26]

A major function of artillery was not so much to win the battle at close quarters, although it could also do that very well, but rather to keep the enemy at bay so he would not come to close quarters at all.

These are primarily defensive functions, and it may certainly be said that the success of Civil War artillery was usually much greater in defence than in the attack. Very long-range overhead fire proved unable to support an attack by searching out the enemy's batteries and regiments[27] unless he left them standing in the open on forward slopes – which in the Civil War, by and large, he did not.

Even in Napoleonic times, however, we find that gunners did better in defence than attack. Most of Napoleon's attempted central break-ins with artillery fell short of the desired effect, whereas most of the defences stiffened with properly concerted mass fires were outstandingly successful. There is therefore little reason to suppose that the Civil War record was really all that very different from the Napoleonic record, or even that the longer-range guns had an important tactical effect. They certainly *would* have such an effect on the battlefields of 1870 in France – but that is an altogether different matter from their effect on the battlefields of 1861–5 in America.

In the Civil War there is little doubt that the importance of artillery lay at short range for cannons, even though that may have been long range for muskets. The Napoleon firing grape and canister was very much a key weapon,[28] which could add a great deal to almost any firing line. It was certainly no accident that, as was noted by one Union soldier in the Atlanta campaign, 'The Rebels seldom fire their cannon except when they can use grape or canister'.[29] In part this was because they were seriously out-gunned at longer ranges, but it was also probably a matter of choice and tactical understanding. It was at canister range that the most decisive results might be expected.

The canister round itself was not the only projectile used at close range, particularly as only four such rounds were statutorily provided in a 12-pounder's ready chest. They could be supplemented at will with unfused shells, spherical case, or even scrap iron.[30] One Confederate officer believed the best ammunition at close range, surprisingly enough, was the distinctly unsubtle and technologically primitive solid shot:

Strange as it may seem, this was much more efficacious for breaking a charging line than shrapnel and canister, which while disabling twice as many did not make such a crashing noise.[31]

Once again, we find that a weapon's utility was proportional to its ability to impress and shock the enemy rather than its ability to kill him. Dead men did not run to the rear spreading panic and demoralisation among those not yet in range of the enemy's guns.

Where there was no artillery to hold the line in the Civil War, an attacker's task was found to be far easier than at the points covered by the guns. At Spotsylvania the success of Hancock's unpromising massed charge was attributed to this factor,[32] while conversely the failures at Cold Harbor were blamed on a deadly cross-fire of artillery. It would thus be a great mistake to discount the value of artillery to a defender, or to assume that it had somehow been superseded by improvements in musketry.

THE ALLEGED VULNERABILITY OF ARTILLERY

Many modern commentators have supposed that the short- and medium-range use of artillery in the Civil War was a logical impossibility – or at best an expression of suicidal bravery – simply because the gunners were vulnerable to the new small-arms fire. It has often been alleged that field guns lost their pre-eminence in battle as soon as the invention of the rifle musket had placed them within range of enemy infantry. According to this analysis the Napoleonic tactic of the artillery charge to canister range had been made obsolete because infantry weapons could at last reach out further than canister could be fired. A battery venturing too close to the firing line could therefore be mown down before its fire took effect: it would be forced to retire to a more circumspect distance if it wished to fight again another day. Hence the only practicable gunfire on the Civil War battlefield is often assumed to have consisted of largely useless pot-shots at ranges above 1,000 yards.

This doctrine is utterly misleading in every way, since not only did Civil War batteries frequently open fire at ranges well below 1,000 yards, and had an important tactical effect at those ranges, but they also suffered relatively few casualties as the price of their temerity. Far from dominating the battlefield, in fact, the unhappy infantry could rarely do very much to stop gunners from moving precisely where they wished[33] or tormenting whomsoever they chose at whatever ranges they cared to select.

It is not uncommon to read of batteries, engaged continuously in hot action for several hours, which 'surprisingly'[34] suffered a negligible number of casualties. At Second Manassas one Union battery, alone and unsupported, even managed to beat off infantry from as near as 65 yards without incurring any loss to itself.[35] Admittedly, it was more normal for losses to rise if the enemy could come so close, but it was very rare indeed for a battery to lose more than 10 per cent of its strength, or an average of approximately two men per gun.

The normal manpower available for each piece was between 17 and 25 men, with the Confederates averaging four pieces per battery and the Federals five. After Fredericksburg the Union authorities wished to outfit each battery with six guns and 150 gunners, but this ideal was never achieved. During the Wilderness campaign they moved down to the Confederate practice of using only four guns in each battery, although on average they did manage to provide ten or twenty more gunners to man them, not to mention enormously superior equipment and gun teams.[36] Even so, it was quite common for both sides to supplement their guncrews with infantry drafts.

Out of all these men, most could expect to survive a battle unharmed, even if the battery was overrun by the enemy – a rare event in itself. For example, at the end of Second Manassas the last battery captured lost only 14 men including 5 prisoners, while earlier Longstreet's men had captured a whole battery but only 34 of its men.[37] At First Manassas a crashing Confederate volley at 70 yards, delivered by troops previously thought to be friendly, captured two Union batteries but hit only 40 men at the very most.[38] At Antietam an Iron Brigade battery suffered 40 per cent casualties, but only after a long fight at close range, unsupported by friendly infantry for much of the time. In the same battle Tompkins's battery suffered 19 casualties in a very hard-fought four hours which included at least one infantry charge to as close as ten yards from the guns.[39] At Fredericksburg two Union batteries were caught in the fighting at the exit from the town and suffered the worst Federal artillery losses of the day – 13 and 16 men, respectively. At Chancellorsville the worst Union battery loss was 46 casualties out of 120 men, although this did not stop that unit firing off its complete supply of ammunition.[40] In military terms we could even claim that the battery's performance was entirely unaffected by the incoming rounds it received.

On a few occasions batteries lost still more heavily than this. Potts at Spotsylvania lost half his men, although once again he did manage to fire all his ammunition. Pegram's battery at Mechanicsville lost 60 per cent, while Purcell's lost 75 per cent in the course of several engagements during the Seven Days. This was a particularly dangerous outfit to join, in fact, since it suffered 200 battle casualties in the course of the war, but such figures were very rare indeed.[41] The normal or average loss to a battery in a major battle stood rather nearer to 5 or 10 per cent of the effectives.

In the four Seven Days battles the Union artillery lost a total of 427 casualties, or 6·7 per cent of its strength, even though the guns had been the backbone of the defence in at least three of the four actions.[42] At Second Manassas the Confederates lost 120 gunners out of something approaching 3,500 – perhaps 3·3 per cent of the total. At Chancellorsville their artillery casualties came to 7 per cent.[43]

At Gettysburg, surely one of the hardest-fought battles of the war, the Union artillery lost about 10 per cent of its strength, and the Confederate artillery 13 per cent, as against a general loss in the battle of about 30 per cent.[44] At Fredericksburg the outgunned Confederates lost 7 or 8 per cent of their gunners against a general loss of only 6·4 per cent, but the Federal artillery lost only about 5 per cent against a general loss of 10·3 per cent.[45] It was quite normal for artillery casualties to be significantly lighter than infantry casualties in these battles.

The figure of 5–10 per cent gunner casualties may be contrasted with an average of at least 14 per cent overall casualties in the twenty-five major battles which I have analysed, and an almost routine acceptance of 30 per cent casualties among those infantry regiments which were heavily engaged. It was not at all unknown for infantry units to lose 60 or 70 per cent, and at Gettysburg 1st Minnesota suffered no less than 82 per cent killed and wounded – a figure which was equalled by three Confederate infantry regiments,[46] but never reached in an artillery unit while it was acting *as* artillery. Ironically the heaviest artillery casualties occurred when the soldiers were taken away from their guns and thrown into the fighting as infantry. The 1st Maine Heavy Artillery had the dubious distinction of suffering the highest losses of any Union regiment in the war, including the heaviest casualties in a single battle. At Petersburg on 18 June 1864 this regiment lost 632 out of 900 engaged – and that just a month after it had lost 476 men at Spotsylvania.[47]

Out of the losses suffered by artillery when fulfilling its normal role, at least as many were likely to be caused by counter-battery artillery fire as by infantry. Hence a battery of 100 men might typically lose 5 men to musketry and 5 to long-range cannonading. To complicate the statistics there might admittedly be a number of horses hit – at least as many again as the human casualties – but we can certainly state that losses in battle to musketry sniping at long range were minimal. When infantry became deadly to artillery it was almost always at the close musketry ranges at which the infantry habitually fought – say between 33 and 100 yards. Such a close approach to artillery remained relatively rare, however, since batteries were usually given as wide a berth as possible. As Colonel Wise explained:

It is very often the case that the greater the influence the guns exert upon the course of a battle, as at Groveton, the fewer the losses the detachments sustain by reason of their ability to hold the enemy at ranges beyond the fire effect of the latter.[48]

In the specific case of Groveton, mentioned by Wise, the main Union attack was channelled into the wooded approaches to the Confederate position precisely because the more open approaches were so thoroughly dominated by Jackson's artillery at medium ranges. Every advance across the open ground went to earth or turned back before it had come into useful musketry range of the guns.[49]

A major exception to the picture painted above is to be found in the use of specialist snipers, especially during sieges. When time was no object an individual marksman with a Kerr or a Whitworth might settle down for an afternoon's harassment of a gun position, from a range beyond the reach of canister.[50] Even a few hits, in these circumstances, might persuade the gunners to retire for a spell, simply because there was no great urgency attached to staying at the front. The operational context was therefore just as important as the technological capabilities of the weapons in determining the final outcome.

Of particular relevance to the operational context, also, was the very important question of whether or not the artillery was supported by infantry. Guns in small numbers exposed to a heavy attack without supports would naturally be much more vulnerable than massed batteries protected by watchful infantry. Thus at Cedar Creek in 1864 an isolated Confederate battery was captured by Union skirmishers, whereas the well emplaced Union gun line exerted an important influence upon the battle.[51] This can scarcely be used as evidence to show that skirmishers had achieved a universally crushing advantage over the guns. Even in Napoleon's time a battery which ventured too close to the enemy without support had been liable to suffer heavy loss, and it is hard to see that the situation had changed in any way by the 1860s.

'IL FAUT S'APPROCHER ET TIRER VITE'[52]

Although the casualties suffered by gunners were normally much lighter than many commentators would have us believe, and the tactical importance of artillery on the defensive was often considerably greater, we must still confront the failure of Civil War artillery to support its own infantry in the attack. Napoleon had supposedly perfected the technique known as the artillery charge, by which guns were manoeuvred close to the enemy line to blow a hole in it, but this does not seem to have been repeated during the American conflict. On the contrary, ineffective long-range fire was usually the only assistance that artillery could offer to an assault, with unfortunate consequences for the latter's chance of success.

The trouble with the artillery charge was that it had never really been perfected even by Napoleon. When he witnessed the most devastating example of the genre, by Sénarmont at Friedland in 1807, the Emperor at first misunderstood what was happening to such an extent that he believed his daring subordinate was deserting to the enemy. Nor was the artillery charge ever really built up into a system which could be called upon at will. In most of Napoleon's battles the preparatory fire of the artillery had been delivered by static batteries at long range and – as in the Civil War – this fire had usually been ineffective.

We probably do not need to say any more than this if we wish to disprove the allegation that artillery had somehow been more useful to an attacker in Napoleonic times than it was to be in the 1860s. The single case of Waterloo should show that the ideal of the artillery charge was observed more often in the dismal breach than in the brilliant execution. Neverthe-less, it was a persistent ideal, and a whole generation of French officers grew up believing in Foy's terse dictum that one must 'get close and shoot fast'. We are entitled to ask why their influence was not more widely felt in the American Civil War.

The answer surely lies in doctrinal ignorance rather than in any other factor. Civil War gunners found it excruciatingly difficult even to in-corporate the basic principles of mass and concentration into their tactics, let alone the more advanced (and risky) concept of deliberately placing guns within close musket range of a well emplaced defender. The very idea was so paradoxical and unlikely that most artillery commanders preferred to fight their war without it. Like Napoleon before them, they just could not see the point.

Some enterprising young officers did, nevertheless, understand that the artillery charge was really the key to victory. They amazed their con-temporaries by following Sénarmont's example, albeit on a small scale, whenever they found an opportunity. Major R.P. Chew was a case in point, commanding horse artillery first for Jackson in the Valley and then for J.E.B. Stuart's cavalry,[53] and beneath him came the rising star – Captain John Pelham. Pelham was hailed by his colleague Haskell as the first officer 'who ever demonstrated that artillery could and should be fought on the musketry line of battle'[54] – or in other words the first inventor of the artillery charge. That was a bit of an exaggeration, but it was true enough that in the American context of the day there were only too few artillerists who shared the same perception.

Unfortunately, however, the brilliant young exponents of bold tactics usually commanded little more than a single battery. At Antietam Pelham had set up thirteen guns in the critical flanking position, but at Fredericksburg he had to repeat the trick with only two pieces.[55] These interventions were enormously successful in their way, but they could scarcely compare in scale to the theoretical possibility of a major counter-move by horse artillery aimed at rupturing an enemy's entire front by a co-ordinated charge.

No large-scale artillery charge ever took place in the Civil War, despite a number of close approximations. Jackson's unmasking of two batteries at White Oak Swamp in 1862 can be seen as an early attempt to emulate Sénarmont, as can the fragmented (but not totally futile) Union efforts to feed guns forward at both Antietam and – less successfully – Fredericksburg.[56] Alexander on the second day of Gettysburg thought he could charge in pursuit of a beaten foe,[57] while the Confederate plan for Malvern Hill is also fascinating because it represents perhaps the most fully developed scheme of the whole war to bring forward a major battery as the preparation for an attack.[58] Its failure doubtless encapsulates the failure of offensive tactics in the war as a whole, but it may be worth remembering that in this particular case the failure was brought about by the enemy's artillery rather than by his infantry.

When a defender was making good use of the ground, and perhaps also strengthening it with fortifications, it was always difficult to exert leverage upon him using the artillery technology of the 1860s. Direct fire at long range was too lightweight, and indirect fire too inaccurate to have an effect. Only a close-range bombardment followed through by a well timed infantry charge could have offered the certainty of success, but that required a degree of co-ordination and flexibility which was beyond the grasp of most artillery commanders. They saw their problem more in terms of breaking free from the infantry,[59] in order to create a massed battery, than in terms of giving close support once the massed battery had been assembled.

Despite this serious conceptual weakness, however, the artillery played an important part in most of the Civil War battles, and was frequently decisive. No army commander could afford to ignore it, or to assume that it had somehow been rendered obsolete by the onward march of small arms. In the Eastern theatre, at least, the trend was all the other way. In the Army of the Potomac McClellan had started by requesting 2·5 guns per thousand men, but the figure was usually nearer 3 per thousand, and on occasion reached 4 per thousand.[60] In the Army of Northern Virginia, despite chronic material shortages of every kind, Lee usually managed to achieve a better ratio of guns to infantry than his Northern opponents. He often fielded 4 guns per thousand men and sometimes reached 7, 9 or even 10 per thousand.[61] Artillery can scarcely have been an irrelevance on the

Civil War battlefield if so shrewd a commander as Robert E. Lee habitually accepted such a disproportionately heavy train of guns in his army.

With Sherman in the West, admittedly, the story was rather different. He appears to have successively lightened his artillery train as the campaign progressed, allowing it to fall from 2 guns per thousand men to just 1. He explained this in terms of the rising quality of his soldiers, saying that fewer guns were needed to give moral support to veterans than to greenhorns. Nevertheless, the terrain and the concept of operations must also have been very important to his decision. A rapid, almost unopposed raid through Georgia gave no opportunities for the massing of large batteries in the grand manner. In the more formal fighting around Chattanooga both sides had maintained an artillery strength well over 2 guns per thousand men, and Thomas was to keep a comparable figure for his operations against Hood at Nashville.[62] Even in the West, it seems, the artillery had established itself as an essential element in combat.

8

Cavalry

*Into the unspeakable jumble in the roadway rode a squadron of cavalry. The faded
yellow of their facings shone bravely. There was a mighty altercation.*

The Red Badge of Courage, p.61

It is not without reason that on the only occasion when cavalry appear in
Stephen Crane's classic novel they are behind the fighting line, helping to
create a traffic jam. Many Civil War soldiers must have felt that this was
really all the cavalry was good for, apart from raiding railroads or hanging
up invalids to get their gold.[1] If we can detect a certain failure to make the
most of either field artillery or the bayonet in the battles of the 1860s, then
we must concede that the record of the cavalry was at least a hundred times
worse.

MISSED OPPORTUNITIES

The participation of cavalry in the major battles of the Civil War was
generally negligible, and we could have found plenty of excuse for leaving
it out of this study altogether. For example, at Antietam – 'America's
Bloodiest Day' – the Union cavalry suffered precisely 28 casualties to all
causes. At Fredericksburg their losses amounted to just 8 men, or less than
one in 1,500 of the total Union casualties.[2] In the whole of the first three
years of the war the cavalry of the Army of the Potomac made only five
mounted charges against infantry in the course of a major battle – very
many less than Marshal Ney's cavalry had made in three hours at
Waterloo.

   The first of these occasions was at Gaines's Mill, where 5th US Cavalry,
250-strong, counter-attacked the Confederates at their moment of greatest
success. The result of this charge, according to the cavalry, was that a
Union reverse was saved from degenerating into a disaster. According to
the Union infantry, however, their own cavalry converted an orderly
retreat into a disorganised rout. It is at least certain that the attack failed to
make any headway at all against the enemy, suffered 150 casualties, and
was all over in a matter of moments.[3]

   Then at the end of the battle of Cedar Mountain the 1st Pennsylvania

179

cavalry charged against five somewhat disordered enemy regiments. It too was repulsed, with 93 casualties out of 164 men.[4] At Chancellorsville yet another holding action was launched, this time with 8th Pennsylvania charging in column of twos down a road flanked by woods full of Jackson's men at the high water mark of their most spectacular victory. This time, surprisingly, there were only 30 casualties out of 400 men – but the repulse was just as total as on earlier occasions.[5]

At the end of the battle of Gettysburg General Kilpatrick ordered two charges, by 1st West Virginia and 1st Vermont cavalry regiments respectively, against the extreme right flank of the Confederate army. These attacks went in across very difficult broken ground against well emplaced Texan infantry, and both were easily repulsed. The total loss in the cavalry brigade concerned is quoted as 98 men during the whole day's combat, including 65 in the charge by 1st Vermont. Hence we may infer that 1st West Virginia lost little more than 25 men in its attack, even though it came within striking distance of the enemy line twice.[6]

In all we can estimate that perhaps 1,400 Union cavalrymen in the East had the experience of charging Rebel infantry in a major battle during the first three years of the war, suffering 365 casualties or approximately a quarter of their strength. Although certain positive results could possibly be claimed for each charge, the overall outcome can only be described as very negative indeed. This finding would seem to suggest that the day of the cavalry charge had passed; that the rifle musket's improved firepower had given a new security to infantry, even if they had not formed squares; and that the American cavalry had really been rather wise not to charge more frequently in these battles than it did, in view of the probable outcome.

A European cavalryman raised in the tradition of Murat and Nansouty would nevertheless probably have wished to point out a number of significant features in these charges, of which the most important is surely their tiny scale. None was as strong as even one full-strength regiment of about 1,000 men, nor even as strong as the very depleted British Light Brigade at Balaclava. If the 'gallant six hundred' on that occasion had lacked the numbers necessary to make an impact, what chance was there for 1st Pennsylvania at Cedar Mountain? The great Napoleon's idea had been to throw in dozens of regiments all together in massed divisions and corps of cavalry – not a squadron at a time.

Our European commentator would probably not have found fault with the timing of these charges, since all of them were launched at the end of a hard day's fighting, either to cover a retreat or to make the first tentative probes of a pursuit. That in itself was quite consistent with classical theory, although what was noticeably missing was any attempt to intervene during the climax of the battle, as a part of the process of deciding winners and losers. Neither the rearguard nor the pursuit functions of cavalry could really hope to decide anything much more than the level of damage once a decision had been reached – so by limiting itself to those functions the cavalry of the Army of the Potomac was effectively abdicating its potentially most important role.

Another significant feature of these charges was that only the first two, at Gaines's Mill and Cedar Run, seem to have taken place on unencumbered ground. The Chancellorsville action took place in a wood which allowed a frontage of only two men to be used – scarcely a legitimate operation of war. At Gettysburg the terrain was slightly more favourable, but not very much more. Our European observer might therefore be permitted to point out that the failure of these three charges probably owed more to obstacles than to firepower, particularly since their casualties were considerably less than those suffered in the first two charges over more open ground.

The conclusion we reach is that no serious attempt was really made to use cavalry in these battles at all, even where the ground was favourable. Such a use was alien to the whole outlook and ethic of the Civil War commanders, and we do not have to look very far to find the reasons.

The number of cavalry available was for a long time derisory by European standards, with the Army of the Potomac rarely fielding more than two regiments per 10,000 men – 8 per cent of the total manpower as compared with Napoleon's 20 or even 25 per cent. Cavalry was enormously expensive to feed, and still more expensive to provide with its essential equipment, notably its horses. A horse cost at least $110, or ten times the monthly pay of a private soldier and five times the price of a rifle musket. On top of this the cavalryman would require a saddle, a sabre, riding boots, a few pistols (preferably the latest Colts) – not to mention horseshoes and tack. The whole proposition was so disproportionately complicated, when compared with the simple needs of an infantryman, that it is scarcely surprising that so few cavalry regiments were put into the field. When we read that there was a cavalry regiment which demanded 81 wagons to support it,[7] at a time when an entire army of 60,000 men might consider 2,000 wagons to be excessive impedimenta, we can begin to glimpse the scale of the problem.

Doctrinal resistance to battle cavalry was also very strong. The 'American tradition' simply had no place for such an animal, since the first US regular cavalry unit had not been authorised until 1792, and the subsequent emphasis had been all upon dragoons or mounted infantry[8] – hybrids which seemed to promise a double return on the government's investment, but which did not really manage to provide either an efficient force of infantry or an efficient battle cavalry. Certainly the influence of Dennis Hart Mahan and his West Point teaching appears to have been in the direction of light cavalry outposts and scouting rather than of heavy combat. It should come as no surprise that Mahan's baleful engineer view of the battlefield managed to eclipse the contribution which a large and energetic cavalry force ought to have made, since we have already seen how influential he had been in shaping the tactical thinking of the infantry. There was a logical spiral at work in this, whereby the greater the emphasis laid by the infantry upon firepower and protection, the less was the attention paid to shock and mobility such as the cavalry could – supremely – provide.

Then again – as with the infantry, as with artillery – the excuse of the terrain was often invoked to show that things could not be done in a European manner. There was some justice in this, especially in the various wildernesses and the Shenandoah valley, not to mention the rolling forests of the West – but in quite a few cases the terrain obstacles consisted of fences or ditches which well trained cavalry should have been able to jump. The problem was really one of training rather than of terrain, since in some accounts of successful charges we hear of fences being thrown down by a regiment's advanced pickets or 'ground scouts',[9] and ditches being leaped at the gallop. At Brandy Station in 1863 the ground had effectively been picked bare by several armies camping there in the course of the previous two years, which showed that cavalry country could be created artificially as well as by nature.[10]

The problem of training, however, remained central. It was not simply that good cavalry took many months longer to build than good infantry, but that the awareness of this fact often deterred Civil War commanders from making the attempt at all. McClellan in particular was notorious for his insistence on long and careful training before his army could take the field, but when it came to his cavalry he seems not to have done anything very much to further this ideal.[11] His awareness of high European standards apparently made him despair that Americans could ever hope to approach them. A whole year was lost as a result, during which the cavalry was allowed to indoctrinate itself with the notion that it could never make use of European methods. In the Army of the Potomac the standards of training and organisation were starting to improve only in the summer of 1863, but by then it was already too late to demand anything more than a hybrid 'American' style of tactics.

The Army of Northern Virginia suffered from a somewhat similar problem, albeit expressed in slightly different ways. Standards of horse care and horsemanship were generally higher, simply because each man had to bring his own horse. One does not need to invoke any theory of 'Southern cavaliers' or 'innate equestrian skills' to see that a soldier will do better if he rides his own cherished four-legged friend than if he is astride an anonymous item of government property.[12] Equally, the principle of massing regiments into brigades and divisions was pursued earlier by the Confederate cavalry than by the Federal, just as it had been by their artillery, due to a greater sense of urgency and military commitment. The well earned reward was a period of two years in which J.E.B. Stuart's horsemen were able, literally, to run rings around their opponents.

The reverse side of these Southern advantages, however, was a shockingly high level of absenteeism and indiscipline. Because the soldiers brought their own horses, they saw themselves as voluntary 'visitors' to the army who accepted neither discipline nor indeed any of the rest of the military contract.[13] They therefore felt free to go home when they felt like it, or to wander away looking for forage, for remounts (often stolen from the Yankees), or for meals (usually from distant cousins resident in the area of operations). This was doubtless a highly agreeable and civilised way to fight a war, but it left the regiments present for duty lamentably weak. The extreme case came in First Tennessee Cavalry, when it unilaterally decided to disband itself completely apart from sixteen men who preferred to stay in camp.[14]

The principle of using cavalry *en masse* was a sound and highly 'European' one, but in the hands of J.E.B. Stuart and his friends it became little more than a licence to roam off into the enemy's rear areas searching for plunder and glory. The success of his early raids in 1862 encouraged Stuart to repeat the process on an ever more grandiose scale thereafter.[15] This led to disaster at Gettysburg, when the cavalry was absent when most needed, although in fairness we can also see that Stuart's example infected the Union cavalry still more disastrously than the Confederate. It was the Northern troopers who were to base a whole alternative system of war upon their enemy's much-publicised exploits, thereby perpetuating the removal of cavalry from the battlefields. One is reminded of the terror bombing offensives in the Second World War, where a temporary German diversion of air effort from the battlefield to city centres (at Rotterdam and then London) led to a massive and permanent diversion of effort, in imitation, on the allied side (from Cologne to Hiroshima).

As the Union cavalry gradually built up its numbers and its skills it came to be used increasingly for raiding rather than for close support of the infantry. At Chancellorsville Stoneman was sent away from the battle with almost all the available horsemen, and the tempo of raiding was also starting to rise in the Western theatre at about the same time.[16] It was to continue to rise right up to the end of the war, which quite appropriately coincided with the culmination of Wilson's Selma campaign – a massive *chevauchée* by 12,000 horsemen supported by 1,500 wagoners and not a single infantryman.[17]

The general Civil War doctrine of raiding, particularly of raiding with cavalry, represented not a great innovation in the art of warfare – as critics such as Liddell Hart have claimed – but its abasement and corruption: a reversion to the methods of the Black Prince rather than a step forward to the *Blitzkrieg* of the twentieth century. The use of cavalry for raiding was not only a deliberate turn away from hope of victory on the battlefield, but it actually removed the means by which victory might have been won at the very moment when those means were at last starting to be properly efficient. If the slow development of high-quality cavalry during the first half of the war was a major missed opportunity, the deliberate diversion of it into raiding during the second half was a still greater one.

REVOLUTIONARY TACTICS

The picture is not entirely bleak, however, since the 'late war' cavalry did eventually develop a style of fighting which gave it, at last, a quite considerable combat power and role in the major battles. Whereas by 1864 the infantry had become largely demoralised and exhausted, the horse soldiers – especially on the Union side – were just then starting to come into their own. They could be invoked as a fresh new force straining to take up the main burden of the war, a spearhead which could use revolutionary new tactics to tackle an old and intransigent problem.

Four major ingredients went into this new tactical mixture: fast operational mobility on horseback out of contact with the enemy; a willingness to take cover and fight on foot when the enemy was close; new repeating carbines to give enhanced firepower; and a mounted reserve ready to make a sabre charge when the moment was ripe. If taken together and used correctly, these four elements could provide important improvements not only upon the cavalry's earlier performance but upon the infantry's too.

What these 'mounted infantry' tactics actually produced was a mixture of fire and manoeuvre equivalent to what Hardee and the *chasseur* school had been trying to give to the infantry. Whereas the *chasseurs* had wished to achieve operational mobility by jogging around the battlefield on foot, the cavalry produced the same result by riding on horseback. Admittedly this was expensive in horseflesh and horse-holders, but it achieved the result more convincingly.

The *chasseurs* also wanted their men to develop heavy firepower with accurate aimed shots from their rifle muskets. In the cavalry the idea was rather to create a heavy barrage of rapid-fire, inaccurate shots, which admittedly made for a rather different concept of tactics; but nevertheless the hope was to create a similar result. Even though the precise means were a little different in the two cases, the intended outcome was identical. The cavalry was simply lucky to have an exceptionally fine weapon placed in its hands, by chance, which moulded the way it used firepower.

Finally the *chasseurs* had hoped to close with the bayonet and destroy a shaken opponent in a mêlée with the *arme blanche*. Despite early doubts and hesitations about this aspect of their role, the cavalry often came to accept it. Typically a regiment might keep back a mounted reserve of perhaps a quarter its numbers, ready to charge when the moment was ripe.

In these circumstances the sabre charge was found to be a useful technique, or in regiments which did not like the sabre the Colt revolver charge had very much the same effect.[18]

These tactics were a mixture between true cavalry action in the mounted charge and 'mounted infantry' action in the preparatory fighting on foot. We might perhaps suggest that it would have been possible to achieve a similar effect in the major battles by attaching a mounted company to each infantry regiment, creating an organic link between the two arms in order to draw the best from each. This had actually been tried in some of the early 'legions' formed in 1861,[19] but as with so many other tactical experiments in the Civil War it was not centrally directed or sustained. The cavalry and the infantry remained two separate and jealously independent services, and they did not normally co-operate as closely as ideal low-level tactics might have demanded.

There were, of course, exceptions to this rule, particularly in small-unit vanguard or rearguard skirmishing. Samuel Barron's last fight in 3rd Texas cavalry seems to have been a model of the kind. It was a rearguard ambush of a Union column marching up a road across a creek. A Confederate infantry regiment lined the creek, concealed by heavy mist. When the enemy came close the infantry opened the action and then the Texan cavalry charged in column from behind, pressing the enemy back two or three miles.[20]

If cavalry more usually had to provide its own 'infantry' to prepare its charges, it did at least enjoy good close co-operation with horse artillery. As we have seen, some of the most daring and aggressive artillerists came from this service, which naturally stressed mobility and opportunism. It was also a dangerous service, since the artillery was often expected to operate in small detachments very close to the enemy's cavalry scouts, where it could easily become involved in skirmishing. Indeed, it was cavalry that generally posed a greater threat to artillery than did infantry, since it could charge to close quarters more rapidly. The large number of guns changing hands at the all-cavalry battle of Brandy Station seem to make the point.[21]

When cavalry was well armed with repeaters and horse artillery, and prepared to fight either on foot or mounted, it could perform all the functions of an all-arms force. In effect it could do without infantry altogether, thus turning on its head the pre-war prediction that 'infantry would make cavalry obsolete'. The main disadvantage with such a force was that it was enormously expensive, requiring a much greater 'tail' – with its foragers and horse-holders – than would otherwise be needed. Cavalry formations could put far fewer men into the firing line than could infantry formations at a comparable level of command,[22] and these men would also tend to feel themselves rather lightly supported. Their forte would lie in staking claims and making rapid assaults rather than in the formal heavy-duty combat of a set-piece battle. In this respect they might be compared with the airborne infantry of more recent times.

Nevertheless, it was precisely in the area of staking claims and making rapid assaults off the line of march that most Civil War armies were deficient. There were few commanders who used their infantry as 'foot cavalry' to hit the enemy at unexpected times and places, to hustle him away from key points before he had settled into his positions. We might even suggest that this was a major reason for the apparent indecisiveness of Civil War battles as a whole, and that if there had been a few more Jacksons to control the march manoeuvres the war would have been more quickly decided.

By the time the cavalry had come to understand its new role as an all-arms force, the need for 'foot cavalry' was chronic. In the Wilderness campaign Grant's infantry had suffered the heartbreaking ordeal of performing a sustained series of rapid flank marches around Lee's right flank, only to find that the Confederates were able to anticipate them and adjust their positions each time. Fatigue and deep mud made it impossible for the Union manoeuvres to reach their objectives before counter-measures were taken, and the result was a futile sequence of parallel battles leading to the siege of Petersburg itself. The failure to move an all-arms force around Lee's flank meant that Grant was condemned to lose heavy casualties for no significant gain. He was forced to give up the attempt and settle down to ten months of siege.

By 1865, however, Sheridan was at hand with a powerful and well indoctrinated cavalry force which could be used as a flanking spearhead of precisely the type the infantry had shown itself unable to provide in 1864. In the Appomattox campaign this cavalry gave a classic demonstration of what could be done.

Credit for the conception must go to Sheridan himself, who resisted Grant's habitual instinct to send the cavalry off on an independent raid. Sheridan insisted that it be kept with the Army of the Potomac and used

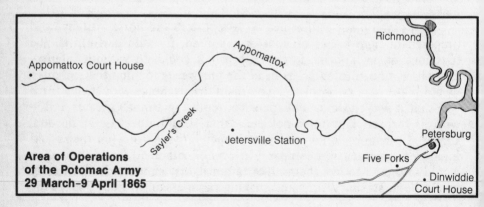

Richmond

Appomattox

Appomattox Court House

Sayler's Creek

Jetersville Station

Five Forks

Petersburg

Dinwiddie Court House

**Area of Operations
of the Potomac Army
29 March–9 April 1865**

for a truly decisive purpose.[23] He saw his role as prolonging the army's left flank, to force Lee to abandon his positions at Petersburg and Richmond, rather than as simply making a sweep in the bad old manner to tear up a few miles of railroad.

Ironically, the operation started badly because the cavalry corps was itself outflanked and surprised by a fast-marching all-arms force under the ill-starred general Pickett. Sheridan suffered a setback at Dinwiddie Court House but was eventually able to stabilise a line and call up reinforcements to crush Pickett's open flank in a fluid encounter battle. These reinforcements were involved in a battle of their own on the White Oak Road, however, and they failed to intervene before Pickett had made good his escape to previously prepared positions at Five Forks. Opponents of Sheridan claim that the whole sequence was an unfortunate diversion from the main business of crushing Lee's army in the White Oak Road battle, where an outright victory could have been won if only the cavalry had not upset the infantry's operations.[24] This may in fact be true, but we must observe that similar infantry operations had failed so often in the past that there was ample justification for the decision to try a different approach on this occasion.

What followed was in any case a magnificent feat of arms by any standards. Sheridan pinned Pickett's men frontally with his cavalry while Warren's infantry executed a most Frederician flank attack. Despite the difficulties of terrain, command liaison and Confederate fieldworks, the position was carried and 6,000 prisoners captured. Admittedly, the cavalry had been repulsed in its attacks until infantry unzipped the position from the side, but it undoubtedly made a great contribution to the final success by its frontal demonstrations and sacrifices. Many of the cavalry units fired off all their ammunition and lost heavy casualties in this action, which was so very damaging to Southern hopes.[25]

Next came a race to Jetersville Station, which the Union cavalry won. Their establishment of a blocking position there was sufficient to head off the Confederates towards the west,[26] and in the ensuing scramble the cavalry once again proved that they could outmarch the enemy's infantry. Custer cut off a third of Lee's shrinking army at Sayler's Creek and, once infantry supports had arrived, captured some 10,000 prisoners. In this battle the cavalry repeated the achievement of Napoleon's cuirassiers at Borodino when they carried a line of enemy breastworks by a mounted charge.[27]

The Confederate cavalry was badly outnumbered and outgunned in this campaign, but it could usually manage to extricate itself from the traps in which its infantry was ensnared. It continued an active and technically impressive resistance right to the end – and it was perhaps no accident that the last man killed in the Army of Northern Virginia was a cavalryman.[28] The end, however, was not now to be long in coming. A new race was won by Custer and Sheridan, this time to the four fat supply trains waiting for Lee at Appomattox Station. The Confederates once again found themselves confronted by a defiant line of Spencer carbines backed by counter-charging mounted troops. Their attempts to break out to the west were again unsuccessful at first, and by the time they had organised a major assault with heavy supports they found that it was too late. The Union cavalry had been stiffened by a strong force of infantry.[29]

The Appomattox campaign does not show us that a battle group of cavalry, even when armed with repeating carbines, could overcome all obstacles. Admittedly, there had been some impressive assaults, both on foot and mounted, in which the rapid fire of the new weapons shocked the opposition and covered the troopers as they advanced. Some commentators have seen this tactic as the forerunner of the twentieth-century concept of 'marching fire'.[30] Yet the rapid-fire weapons also ran out of ammunition more rapidly than the slower rifle muskets, and Sheridan's men were repeatedly being told not to waste their rounds.[31] Conversely, the new carbines suffered from various technical teething problems, especially in the rimfire ammunition itself. The volume of fire which could be developed in action often fell far short of expectations. For example, in one intense firefight in May 1864, 1st Pennsylvania cavalry managed to fire only 12–18 rounds per man per hour,[32] which was theoretically the volume of fire they could produce in one minute.

In the Appomattox campaign there were some failures which show us that mounted infantry was not the complete 'panacea weapon' that its champions would like us to believe. At Dinwiddie Court House Sheridan was badly mauled by Pickett's shrewd blow, and at Five Forks the cavalry found it could make no headway against fieldworks. At Jetersville, Sayler's Creek and Appomattox itself the cavalry's resistance would probably have been overcome if infantry reinforcements had not arrived in time. The possession of technically advanced small arms did not ensure victory against all odds.

What the Appomattox campaign does show us, however, is that cavalry could add considerable extra zest and impetus to the normal operations of infantry. The high mobility of horsemen allowed them an extra freedom on the battlefield which could unsettle any Civil War commander accustomed only to the sedate evolutions of foot soldiers. Even if they did not arrive at the key point with the 'mostest' combat power, the cavalry could at least get there 'fustest'. This made a magnificent multiplier of the infantry's natural force, so that when used in conjunction the two arms became far more formidable than either had been on its own. It is greatly to Sheridan's credit that he understood this and pressed his ideas upon his more traditionally-minded superiors. It is greatly to the discredit of many other Civil War commanders that it took so long for the all-arms battle-group to enter their tactical thinking.

CONCLUSION

The Last
Napoleonic War

*He found that he could look back upon the brass and bombast of his earlier gospels
and see them truly. He was gleeful when he discovered that he now despised them.*
The Red Badge of Courage, p. 109

We are now in a position to look back at the 'brass and bombast' of the
various historical explanations put forward to explain the indecisiveness of
battle in the Civil War. Let us review them here in turn.

First there is the idea that the rifle musket revolutionised tactics. This is
demonstrably false, simply by reference to the short range and long
duration of the firefights. Attacking formations were not mown down
before they arrived close to the enemy, but managed to install themselves
at highly 'Napoleonic' ranges and slog it out until all ammunition was used
up. Nor did the new rifles prevent artillery from operating at canister
range from the enemy. On the contrary, the gunners continued to play the
important part they had always played, and suffered relatively low
casualties even when they were overrun. Nor was the cavalry swept from
the field by improved firepower. It had hesitated to venture onto the field
during the early years as a result of unsound doctrinal preparation, but by
the end of the war it had actually become a decisive weapon, even when
using the sabre.

Improved weaponry did not force the armies to dig fortifications, but
fashion and book-learning did. Once the fieldworks had been dug they
became symbols of specific tactical qualities – firepower and protection –
which Civil War soldiers had decided were most important. As the
fortification principle was extended, the alternative tactics of mobility and
shock fell proportionately into disrepute. By 1864, when the armies had
passed their peak of enthusiasm and energy, both sides quietly but
determinedly boycotted the *attaque à outrance* in favour of indecisive
stand-off skirmish action which saved lives even if it did not win the war.

Shock tactics lay at the root of Hardee's drill manual and were well
understood by a minority of Civil War officers. On occasion they could be
brilliantly successful, and it would certainly be a mistake to underestimate
their potential for winning battles. Unfortunately, however, they were not

properly disseminated among the improvised young armies, nor was any sustained analysis of tactical method undertaken by central authorities. The choice of tactics was left in the hands of individual colonels and brigadiers, which meant that performance remained haphazard and patchy. Where innovations such as the Indian rush or trench infiltration were applied, they could be little more than local and isolated cases.

In these circumstances most regiments which came under the enemy's close-range fire followed their natural instincts and settled down to fire back. Even though the fire might be capable of hitting only one or two men in the regiment during each minute, the imposing aspect of the enemy's position was sufficiently unnerving to deter the attacker from pressing home his assault. The firefight dragged on until exhaustion set in or nightfall put an end to hostilities. Casualties mounted because the contest went on so long, not because the fire was particularly deadly.

Other mistakes were also made, especially persistence with excessively deep assault formations. These had been discredited in Napoleonic times and superseded by more modern tactical concepts. Nevertheless, the Civil War commanders often continued to use them without thinking through such implications as the need for effective passage of lines. The result was often a chaotic intermingling of regiments under the enemy's guns, and hence a loss of impetus. Worse still, the survivors tended to blame all pre-war tactics indiscriminately for the failure. The forward-thinking methods of Hardee and the *chasseurs à pied* were condemned, ironically, for the very faults which they had tried to avoid.

Admittedly, the new tactics of the 1850s were very demanding in terms of both physical effort and military *esprit*. They could not always be made to work even in France, the land of their birth, and it would have been unrealistic to expect raw militia armies to master them overnight. Nevertheless, the raw militia armies of 1861 had shown by the spring of 1862 that they could fight hard and well, and that they could often perform quite complex drill movements quite near to the enemy. The better regiments could therefore quite easily have been led through more advanced tactical training designed to strengthen the weak points in Hardee. The result could have been a corps of shock troops designed to convert one of the many successful attacks into a decisive victory.

Above the level of minor tactics we find various other explanations offered for the indecisiveness of the battles – for example, 'the terrain interfered with command and control'; 'the tactical articulation of nineteenth-century armies favoured a defender rather than an attacker'; 'the logistic difficulties hindered pursuit'. There is doubtless a grain of truth in each of these theories, but there are counter-arguments too. We cannot help noticing a certain breakdown of doctrine on many occasions when a battle of annihilation might have been possible. The Appomattox campaign of 1865 stands as a brilliant example of what might have been achieved in earlier years, if only a proper *corps de chasse* had been available. It stands as a reproach to all the generals who fumbled their encirclements and pursuits, and reminds us that Civil War commanders probably enjoyed just as good means of controlling their battles as had their Napoleonic predecessors – but felt less sure of themselves when it came to higher battle-handling. Faulty understanding of the use of army reserves, in particular, can be seen as a characteristic of the 1790s battles which Napoleonic experience had overcome.

At the end of the day, however, the indecisive outcome of so many Civil War battles must be put down to the individual personalities of the generals. Many of them seem to have been cautious men, rather too ready to call off their battle when victory was in sight, or to refuse combat altogether if the odds were not to their liking. Despite the Lees and the Jacksons, the Hoods and the Burnsides, the predominant military culture was not deeply rooted in rapid manoeuvre and crushing assault. A more tentative and sedate style of war seems to have been more generally preferred.

In many respects the Civil War was comparable in scale to the European warfare of almost any four-year period of Napoleon's career. The armies were roughly the same size and the distances covered were comparable. Admittedly, some of the Civil War operations would have been extremely difficult without the assistance of steam transportation – particularly the inland campaigns in the Western theatre. Nevertheless, the major campaigns in the East could all have been sustained by horse power and wind power if there had been no steam. The battles were certainly fought with essentially Napoleonic weaponry and tactics, although the doctrinal emphasis on fortification would doubtless have appeared a little odd and old-fashioned to the restless Corsican. He would presumably have been mightily puzzled to discover how such an approach had been identified by West Point professors as being the logical lesson of Napoleonic warfare.

A few significant military factors had unquestionably changed since Napoleonic times, especially the ability to switch reinforcements quickly from one theatre to another by rail. Taken together with the new strategic telegraph, this represented a genuine innovation in the art of war, although its effect was often counterbalanced by other factors to the extent that it scarcely seemed to change the overall tempo of operations. In naval affairs the technological revolution may justly be seen as fundamental, but on land this claim seems somewhat exaggerated.

If we wish to find land combats which are truly the first modern battles in history, it is surely to the north-east corner of France in the year 1870 that we must look. Here we find march manoeuvres integrated with a rail-born mobilisation such as was scarcely glimpsed in the America of Davis and Lincoln. We find effective massed shelling at ranges significantly greater than the 1,000 yards of classical practice, and we find the infantry going to ground many hundreds of yards from their opponents, under genuine long-range musketry and – Dr Gatling's early experiments notwithstanding – the first large-scale use of machine guns. In the Franco-Prussian War the new weapon technologies which had been little more than blueprints and prototypes in the American Civil War finally came forward as robust, practical implements for everyday use. Only the survival of successful massed battle cavalry seems anachronistic to the modern observer, although its many achievements in 1870 can at least help to underline what might have been in the earlier conflict.

In political and ethical terms the Civil War seems almost gentlemanly and restrained when set beside the cynical amorality of the Franco-Prussian War. Bismarck's Realpolitik had a calculated quality about it which the passionate Sherman could scarcely rival, while the cold-blooded execution of 25,000 communards within a few days constituted an atrocity far more shocking than the casual abuse of prisoners in Point Lookout or Andersonville over a period of years.

Perhaps it is only in one particular respect that the Franco-Prussian War looked more Napoleonic than the Civil War, namely in its short duration and decisive result. The attacker, who also had the big battalions, simply rolled over his opponent and won. Such a cut-and-dried victory seems strangely old-fashioned to us today, accustomed as we are to conflicts in which greatly superior war machines are stalled and frustrated by the resistance of weaker opponents. Whether we look at the two world wars or the operations which followed 1945, the latter seems to be a constant expectation in twentieth-century warfare – just as it was a feature of the Civil War.

The duration and frustration of those indecisive operations in the 1860s should not, however, be seen as an inevitable result of a deep change in the nature of warfare. On the contrary, there were many turning-points at which different decisions might have led to a much quicker end to hostilities, in some cases even to a Confederate victory. If certain winning positions had been seized and exploited, if certain training programmes had been pushed forward and certain tactical doctrines avoided, if staff work had been centrally organised and shock troops selected, if the engineers had not exercised a stultifying influence on military thought – then the Civil War could have shown us decisive results as dazzling as anything seen in the days of Napoleon.

APPENDIX I

The Art of
Tactical Snippeting

The true tactical historian is such a rare and little-known animal, even among military historians, that it may be worth while to pause for a moment to consider his role and methods of working.

Traditionally, the tactical historian would decide which set of battles he wanted to study and then start by looking into their weapons and drill manuals. These two items of hardware would give him a set of basic statistics upon which he could base a theoretical model. The weapons would yield data on accuracy, rates of fire and range; the drill books would explain the various troop layouts available and the speeds at which each one could move.

If all this was brought together, the historian could estimate the number of projectiles which could hit any given attacking formation before it reached a defender, and vice versa. By gaming through the various combinations of drill layouts in attack or defence the relative value of each could be discovered. From there it was but a short step to awarding praise or blame to historical commanders according to their choice of formations in particular battles. Thus the French could be given low marks for their use of columns at Waterloo, but the British given high marks for their firepower from a deployed line. The Prussians of 1870 could be given higher marks than their French opponents for preferring company columns to battalion columns – and so on. A whole new field of geometry was opened up, which promised to explain the inner causes of victory and defeat, and hence the rise and fall of nations.

Alongside this 'geometrical' or 'hardware' approach there have always been 'literary' or 'subjective' treatments of men in battle – works like *The Red Badge of Courage* itself. In quite recent times, however, we have seen a new strain of writers emerging from this genre and posing a more direct challenge to traditional tactical history. Let us call the new breed 'compassionate military historians' – writers who have no time for mathematical hocus-pocus, simulation theory or military manuals of any sort, but who none the less purport to be practitioners of the science of tactical history. Such writers tend to be longer on inspired generalisation and empathy for doomed youth than they are on careful study of the sources. They admittedly help us to bring our rich heritage of military memoirs and personal impressions more centrally into the picture – which is a welcome corrective to the earlier tendency – but it is not always clear that they can take us very much further than that.

From today's perspective we can see that neither the 'hardware' nor the 'compassionate' lobby on its own can give us good tactical history. The older-established school was too ready to regard soldiers merely as pawns in a game of advanced mathematical chess, and had no notion of what actually took place when human nature intervened to upset the theoretical calculus. Such writers could not conceive that a rifle musket sighted up to 1,000 yards was effective, in the hands of real soldiers, only up to about 33 yards. Conversely, the more modern school knows a great deal about how soldiers feel, but does not bother too much with the physical and doctrinal matrix in which they work. Like the Kaiser in the First World War, our 'compassionate' military historian will call for anecdotes and 'trench stories' rather than solid operational analysis.

The worst tactical history of all is the sort which combines the vices of both sides and tries to base the dehumanised logic of game theory upon the unsystematic anecdotage of literary insight. We must avoid such sloppy thinking and try, instead, to achieve a judicious combination of whatever is best in each of the two approaches. From the 'compassionate' school we can draw a deeper understanding of the psychology of combat, as well as a more effective use of military memoirs and regimental histories to illustrate what actually took place. This can teach us – not before time – to be infinitely suspicious of the assumptions which a study of the hardware can generate. Yet on the other side we must also remember the realities upon which the 'hardware' school is based. Drill *did* have a certain value, even if it was not quite what Hardee might have intended; and Boston office boys *could* sometimes hit their man with an Enfield rifle, regardless of the stresses and strains of real combat. Even the despised wargamer's view of military history can sometimes teach us important lessons which its literary denigrators will never begin to discover. Denial of any one of these many different approaches to military analysis can be quite as misleading as obsessive immersion in any one of them.

The tactical historian should have a wide variety of techniques at his command, but most of all he must look at autobiographical memoirs, diaries, letters from the front and regimental histories written by contemporaries. It is through this type of literature that the most urgent voices from the past can speak to us, and if properly understood they can give us a far more immediate impression of the battlefield than any number of drill manuals or technical treatises. What the participants retained in their minds strongly enough to wish to tell us is surely precisely the sort of thing which we ought to be dissecting most carefully. Hence we should not try to ask 1986 questions of our 1862 witnesses until we have at least come to terms with what they were trying to say in the language of

1862. For example, we should not assume that their talk of a 'bayonet charge' was necessarily stupid or self-serving simply because we believe the bayonet to be an obsolete weapon in twentieth-century warfare. Nor should we dismiss their enthusiasm for the assault as a subjective anomaly, just because we think we know that assaults always received a bloody repulse. The question we should be asking is not so much 'Was this opinion right or was it wrong?' but 'Why did these men in the 1860s hold the opinions they did?' It is for the scientist to decide who was right and for the littérateur to mourn the others; but it is the duty of the historian to discover the hows and whys of what really happened.

A major difficulty arises from the fact that most volumes of memoirs and regimental history contain only a few short passages dealing with battlefield tactics, and some become systematically reticent precisely at this point. The tactical historian must therefore tread warily and be prepared to pounce upon even a fleeting and ambiguous reference to what was going on. This 'tactical snippet' may then be added to his collection of evidence and, once it has been compared with the rest, may yield deeper levels of meaning than was at first suspected. For example, a reference to discarding packs before a charge may not at first appear significant, but once half a dozen such references have been assembled we begin to see that there was an art and science – indeed a veritable ideology – of discarding packs. This simple gesture contained a wealth of inner meaning which the casual reader in the 1980s may all too easily overlook.

No turn of phrase employed in a tactical snippet should be ignored, since it may conceal an unexpected or veiled implication. There is probably a need for detailed content analysis of these documents along the lines used by Michael Barton in his study of the social backgrounds of Civil War soldiers but, unlike him, taking the specialised vocabulary of combat as a base. The drill manuals do not always use English in the same way as Webster's dictionary.

A single snippet no longer than a paragraph may quite often yield material for many pages of comparison, speculation, interpretation and exegesis. It is therefore the duty of the tactical historian to practise and develop his skills in this direction as frequently as he can, even though only a very small part of his findings may eventually find their way into print. Private correspondence or informal discussion in the columns of amateur journals can be an invaluable adjunct to this process, since two heads are often – admittedly not always – better than one. We should certainly reject the overbearing scorn towards the amateur military history press which, alas, is exhibited only too frequently in academic circles. Fortunately for the tactical historian, however, the many flourishing amateur journals appear to be in no danger of disappearing.

The art of tactical snipeting really forms the essential homework which a serious student must do before he can get to grips with the heart of his subject. It is not as widely practised as it might be, and too many tactical historians try to get by on airy generalisations which are demonstrably false. But as the study of the battlefield gradually grows in stature as a respectable subject, so we may be sure to see a corresponding growth in the art of tactical snippeting.

Ardant du Picq was perhaps the first true tactical snippeter in his *Battle Studies*, but interesting recent examples include Steve Fratt's *American Civil War Tactics* and (for the Napoleonic period) the work of Jim Arnold, Ned Zuparko and John Koontz for *Empires, Eagles and Lions* magazine.

APPENDIX II

The Decisiveness of Civil War Battles

Several recent works have made elaborate attempts to quantify tactical effectiveness on the basis of *casualty* statistics from the Civil War battles. What has not been done, however, is to analyse statistics of *military results* as compared with those in other eras of military history. An attempt to fill this gap is presented here.

Four sample sets of battles have been chosen – Napoleon's major battles of 1800–15; the major battles of the Western Front of 1914–18; Livermore's Civil War battles of 1861–3 in which 1,000 or more casualties were suffered by either side; and my own selection of the major battles from the whole of the Civil War. The statistics are given in Table A II.1. I have included the Livermore list because it has been used as the basis for studies by other writers and therefore makes an interesting comparison, whereas in the other three cases the lists have a somewhat less scientific basis but are more representative of what we think of as the major battles of the periods in question.

TABLE AII.1   MILITARY RESULTS OF SAMPLE BATTLES, 1800–1918

	Napoleon, 1800–15	Western Front, 1914–18	Liver- more's list, 1861–3	All Civil War
Total in sample	25	25	26	25
Tactical victories won by tactical attacker	12 (48%)	11 (44%)	9 (35%)	9 (36%)
Strategic victories won by tactical attacker	15 (60%)	9 (36%)	12 (46%)	14 (56%)
Strategic victories won by strategic attacker	14 (56%)	8 (32%)	11 (42%)	11 (44%)
Average duration in days	1·6	32	2·1	2·4
'Pyrrhic victories'	6 (24%)	16 (64%)	7 (26%)	11 (44%)
'Outstandingly decisive' victories	7 (27%)	2 (8%)	4 (15%)	5 (20%)

These figures suggest that even in Napoleonic times the tactical defensive was marginally more successful than the offensive. It was, after all, the Napoleonic experience that led Clausewitz to conclude that 'the defensive is the stronger form of war'. Even more surprising, perhaps, is a large number of First World War offensives that achieved their tactical objectives, regardless of whatever else they did or did not achieve. Admittedly, these occurred mostly in the fluid first two months of the war, or in the equally fluid last nine months; but they still help to highlight the military logic which continued to produce battles under conditions which many observers believed should have made battle itself obsolete.

The Civil War record for tactical success in the tactical offensive is the worst of all, but it still includes over a third of the battles fought. If we remember that the average loss of life in Civil War battles was lower than in either the Napoleonic or Western Front examples, we can perhaps speculate that Civil War engagements tended to be called off prematurely, before the cost had mounted too high – but also before the attacker's objective had been achieved.

This suggestion gains support from the number of strategic victories achieved in the Civil War by attacks which had tactically failed. This would seem to indicate that Civil War commanders paid rather less attention than Europeans to the actual battlefield results, but had leisure to choose their strategic posture according to their general feelings. Hooker at Chancellorsville and Bragg at Perryville spring to mind as examples of this.

However that may be, it is worth noting that both the Napoleonic Wars and the Civil War showed a tendency for failed attacks to be converted into strategic successes, whereas the First World War seemed to go the other way. Whatever had changed in the conditions of warfare by 1914 tended to work against a general's ability to use a local victory for wider purposes. Hence the overall efficacy of First World War attacks was almost half that of Napoleonic ones. When viewed in this light the Civil War attacks seem to be almost as effective as Napoleonic ones, despite their poor tactical showing. If we express this in gambling terms, we could say that Napoleon knew he had a two in three chance of ending on top if he launched a tactical attack; Foch and Haig had a chance of one in three, while Grant and Lee had evens.

Something similar can be said when we consider the relationship between strategic offensives and strategic success. The conditions of Napoleonic warfare seem to have been most favourable to the strategic attacker, those of the First World War least favourable. The Civil War apparently comes in somewhere half-way between the two, but still with little less than an evens chance of victory for the offensive.

So far the military results seem to indicate that the conditions of Civil War fighting offered rather more success to an attacker than the popular stereotype might suggest. As far as the *perceived* decisiveness of these battles is concerned, however, we can agree that by the time of the Civil War there had indeed been a decline from Napoleonic standards. The battles took longer – perhaps half as long again as the average Napoleonic action, and then normally with fewer men engaged. There were more apparently 'Pyrrhic victories' in which the margin of success appeared slight by comparison with the heavy costs. Whereas Livermore's list for the first half of the war shows almost a Napoleonic level of such distressing events, the list for the whole war includes considerably more. It would seem that the generals (if not the men) had become more inured to the shedding of blood by 1864 than they had been earlier in the war. Finally the Civil War had fewer 'outstandingly decisive' victories in which an enemy army was destroyed or a peace treaty forced. Admittedly, Napoleon's record seems quite exceptional in this respect, since he encountered no less than seven separate coalitions and fought each of them to a decisive conclusion, one way or the other. Marengo and Austerlitz look rather less decisive, however, if we remember that the 'decisively beaten' Austrians returned to the fray four years later in each case.

The First World War is clearly the odd one out when it comes to perceived decisiveness, since its average battle lasted a month and was very definitely 'Pyrrhic'. The only two 'outstanding victories' that spring to mind for the whole four years of the Western Front are the German offensive through northern France in the first six weeks of the war and the inter-allied attack at Amiens in August 1918. Both of these seemed to achieve rather more profound and lasting results than any of the other successful attacks, although neither of them in themselves achieved the destruction of an army or the conclusion of peace.

Were it not for the single fact that Civil War assaults tended to be tactically less successful than those of other wars, we would have little difficulty in assimilating the Civil War with the Napoleonic experience rather than with the First World War. As it is, however, the failures on the battlefield have clouded the issue and led many commentators to the opposite conclusion. Yet again we find that it is the minor tactics of a war that have a determining influence on the way in which we think about it.

Some recent statistical analyses may be found in Dupuy (mostly twentieth-century battles, but the methodology could be transferred to the Civil War); Kirkpatrick; McWhiney and Jamieson; and Hattaway and Jones. The battles I have used in the present survey are the following:

Napoleonic Arcis, Aspern, Austerlitz, Bautzen, Berezina, Borodino, Craonne, Dresden, Eckmuhl, Eylau, Friedland, Jena, Laon, Leipzig, Ligny, Lutzen, Maloyaroslavetz, Marengo, Montmirail, Quatre Bras, La Rothière, Smolensk, Ulm, Wagram, Waterloo.

Western Front Amiens, Arras, Cambrai, Champagne 1915, Chateau-Thierry, Loos, Lorraine 1914, Ludendorff's 1918 offensives I–IV, Marne I and II, Meuse/Argonne, Mons/Le Cateau, Neuve-Chapelle, Nivelle offensive, St-Mihiel, Somme, Verdun (German and French offensives), Vimy, Ypres I–III.

Livermore's List Antietam, Arkansas Post, Cedar Mountain, Champion's Hill, Chancellorsville, Chattanooga, Chickamauga, Chickasaw Bayou, Corinth, Fort Donelson, Fredericksburg, Gettysburg, First and Second Manassas, Mine Run, Murfreesboro, Pea Ridge, Perryville, Prairie Grove, Richmond, Seven Days, Seven Pines, Shiloh, South Mountain, Williamsburg, Wilson's Creek.

All Civil War Antietam, Bentonville, Chancellorsville, Chattanooga, Chickamauga, Cold Harbor, The Crater, Five Forks-Appomattox, Fort Donelson, Franklin, Fredericksburg, Gettysburg, Kenesaw Mountain, First and Second Manassas, Murfreesboro, Nashville, Peachtree Creek, Perryville, Seven Days, Seven Pines, Shiloh, Spotsylvania, Weldon Railroad, Wilderness.

NOTES

Most of the works cited in these notes are listed in the bibliography. When they are, I have referred to them here simply by their author's name or the abbreviated title.

PREFACE

1   I am encouraged by the thought that at least two British writers have left an indelible mark on the Civil War debate – G. F. R. Henderson and J. F. C. Fuller. Although I cannot hope to emulate their achievements, I have at least been privileged to consult items from their libraries in the Camberley Staff College.
2   See Jones, pp. 109, 224.
3   A similar motivation appears to have lain behind the writing of the excellent Civil War history by Hattaway and Jones, while even more specific is James Reston jr's *Sherman's March and Vietnam* (Macmillan, New York, 1985). For a succinct and stimulating analysis of the American military mind it is hard to beat the article by Shy, although he sees the Civil War more as a reinforcement of existing thought patterns than as a blueprint for new thinking.
4   The edition of Crane I have used gives invaluable commentaries on his sources and the great influence he exerted upon such authors as Conrad and H. G. Wells. We might surely extend this list to include many other twentieth-century war writers, since Crane had such a truly seminal role.
   An interesting discussion of 'seeing the elephant' in the Civil War may be found in Wiley, *Billy Yank*, p. 69.
5   See my *Forward Into Battle*.
6   The *Empires, Eagles and Lions* magazine, edited by Jean Lochet, is an honourable and often highly scholarly exception to the general rule of Napoleonic apathy, although the scarcity and unreliability of sources for that era is a persistent theme in its pages.
7   Privately produced copies of the transactions of this conference are available through the author, covering many military and naval aspects of the Civil War.

INTRODUCTION: THE ALLEGED ORIGINS OF MODERN BATTLE

1   De Forest, p. 62.
2   Crane, p. 89.
3   See Appendix II for the actual outcomes of Civil War battles.
4   The debate after 1865 is well examined in Jamieson, p. 182ff, although by its nature the views of those who saw no revolution since Waterloo were presented less forcibly than those of believers in a dramatic change.
5   McWhiney and Jamieson, p. 6, and Mahon, p. 68.
6   McWhiney and Jamieson, p. 7.
7   Mahon, p. 59.
8   Hattaway and Jones, p. 720. My emphasis added.
9   Often wrongly attributed to Sheridan, this remark was actually Grant's – Worsham, p. 178.

10   For historiographical aspects, consult Johnson, Pressly and Leonard.

11   Chamberlain, p. 267, generally confirms this atmosphere at Appomattox, but notes that the Confederate General Wise did not share it. 'We hate you, sir' was *his* conciliatory greeting.

12   Modern work on Civil War photographs, notably by William Frassanito, has shown how many of the most haunting scenes were carefully posed for maximum effect. But they remain haunting all the same.

13   I have used Chandler for Napoleonic casualty statistics. Fox, p. 47, makes a case that rather lower figures applied at both Borodino and Leipzig, and is at pains to show that the percentage lost out of each army was lower in European battles than in those of the Civil War. Nevertheless, the main point remains that the Europeans lost larger numbers of men in a shorter space of time.

14   These issues are discussed in, e.g., Philip Knightly's *The First Casualty* (Deutsch, London, 1975) and – especially – Paul Fussell's *The Great War and Modern Memory* (OUP, London, 1975).

15   Mitchell, p. 207; Hattaway and Jones, p. 692.

16   Leonard, p. 9.

17   Edwards, p. 434, picture caption.

18   Bull, p. vii.

19   Harsh, p. 63, sees the Civil War as the 'First Modern War' for a different reason – the scale of the human mobilisation. This view is also followed in John Terraine's *The Smoke and the Fire* (London, 1980), although it is tempting to point out that the French Revolutionary *levée en masse* had been called some seventy years before the 1860s.

20   Hattaway and Jones, p. 47–8.

21   Nisbet, p. 68.

22   Luvaas, p. 126.

23   Hattaway and Jones, p. 538.

24   Moseley, p. 14.

25   Morrison, *passim*; Hattaway and Jones, pp. 173, 288; McWhiney and Jamieson, p. 18.

26   Haskell, p. 55–6.

27   ibid., p. 56. Compare Shannon, 1, pp. 153–65, for a staunch defence of West Point.

28   The nature of Southern society is discussed by Michael Adams. Certainly the memoirs of Confederate staff officers (e.g. Douglas, Haskell, Sorrel) portray a very close aristocratic circle, in which all the leading military figures seem to be related to each other.

29   McWhiney, p. 262.

30   Bull, p. 164.

31   The debate about the influence of Jomini has been especially seriously tainted by this line of thinking, and we are sometimes almost told that the medieval brutalism of a Sherman or a Wilson was a higher form of war than the more elegant strategies of the Swiss author. Burning undefended towns is more 'political' than outmanoeuvring enemy armies, apparently.

Two points are often missed about Jomini. One is that whatever we think of his dubious professional morality (as both a soldier and a writer), we have to admit that he did very well summarise the highest military thinking of his day as far as reasonably small armies were concerned. The fact that he did not discuss either *der Millionenkrieg* or the political application of violence may have been unfortunate for the French in 1870, but it can scarcely be blamed for any of the disasters of the Civil War in America. Scarcely any army in that conflict was larger than 100,000 men, while the political issues were perfectly plain for all to see.

Secondly, Jomini cannot be held to have exerted any great influence on the mass of West Pointers, even assuming they read him at all, simply because their syllabus was so crammed with other things which were of greater immediate concern. West Point was a school for second lieutenants, not for generals. Such schools are better at teaching the minutiae of drill or weaponry than the finer points of *Weltstrategie*. Admittedly, a few enthusiasts such as Halleck and Beauregard did take Jomini seriously, but even with them his message seems to have been distorted by the engineers. It would probably be fair to suggest that the West Point ideal of a French general looked less like Jomini than like Vauban wearing Napoleon's hat.

The dispiriting recent debate about Jomini is summarised in Hattaway and Jones, pp. 21-4. A splendid modern account of Jomini's work is John Shy's *Jomini* in Peter Paret, ed. *Makers of Modern Strategy* (Oxford University Press, 1982; pp. 143-185).

32 Ward, pp. 13 and 15, notes that troops lacking rubber blankets had priority for any roofed accommodation. In general for these items of equipment see Coggins, both volumes by Francis Lord, Todd, and Billings.

33 Black, p. 32; Coe, pp. 71-2; Green, p. 78-9; Patrick, p. 34; Barron, p. 82.

34 See English, *Confederate Field Communications*, Fishel and – especially – Thompson.

35 General descriptions of weaponry may be found in Coggins, Francis Lord and Edwards. General 'Beast' Butler was perhaps the most tireless inventor of secret weapons in the war – see Eisenschiml and Newman, 1, p. 594.

36 Sorrel, p. 67, *Battles and Leaders*, 2, p. 201. Ground torpedoes were also used at Charleston, 1863 (Ward, pp. 48, 56) and by the Union at Petersburg in 1864 (George Wise, p. 212).

37 Sorrel, p. 214; *Thirty-fifth Massachusetts*, p. 193; *Battles and Leaders*, 3, pp. 741, 749.

38 *Battles and Leaders*, 4, pp. 202-3. Abner Doubleday had thought of wire entanglements in his preparations to defend Fort Moultrie before war broke out, but it was never put to the test (*Battles and Leaders*, 1, p. 43). In none of these cases was the wire barbed.

39 The Confederate General Alexander was an early champion of this view – see Moseley, p. 142. Shannon, 1, pp. 107ff, restated it forcibly in the 1920s but has been debunked recently by Carl Davis.

40 Edwards, pp. 149-50, 159-163.

41 These propaganda points are discussed at length in Berger.

42 Carl Davis, p. 66.

43 ibid., p. 164.

44 Starr, in *Cold Steel*, discusses this, and shows that the sabre was used effectively more often than is commonly imagined. For Mosby's complete contempt for the sabre see Wilson, pp. 307-29.

45 The fullest presentation of these statistics is in Moseley, pp. 195-212, although they are also well known to, e.g. Mahon, Buechler and George Adams. Essentially they show that in the spring of 1864 the Army of the Potomac's wounded included about 6 per cent hit by artillery shells or roundshot, 0.1 per cent by bayonets and 0.05 per cent by swords. Self-inflicted wounds (presumably including accidental) ran at around 7 per cent, or a similar proportion as to enemy artillery. All the rest were to enemy musketry.

46 Moseley, p. 193. For bowie knives see, e.g., Jones, p. 3, Shannon, 1, p. 41; for machetes, e.g., Barron, p. 28; for pike manufacturing, e.g., Green, p. 19; and for cutlass drill for storming entrenchments as late as 1863, see Ward, p. 79.

47 Lewis, p. 85.

48 Numerous writers with rifle expertise had predicted the 'rifle revolution' for decades before the Civil War – e.g., Colonel Marnier's 1837 pamphlet for the French army, recommending rifles as 'hand artillery', *Améliorations proposées dans l'armement et l'éducation des troupes* (French army archives Vincennes, MR2140); or Panot's *Cours sur les Armes à Feu Portatives* (4th edn, Paris, 1851). Busk in Britain was ideologically committed to the rifle volunteers, while Cadmus M. Wilcox in the USA was also predicting the imminent demise of artillery in his *Rifles and Rifle Practice* (New York, 1859, quoted in Jamieson, pp. 50, 53, 56). As Richard Munday has shrewdly pointed out, however, the existence of this propaganda before the war should not be confused with what happened during it, even though over hasty historians are too often prepared to do just that.

Notes

CHAPTER 1: THE ARMIES LEARN TO FIGHT

1 See Appendix II.

2 There is a vast wealth of good general histories of the war, and I have been able to read only a small fraction of them. Of these, I have particularly appreciated Mitchell as a starter and Hattaway and Jones as a solid military analysis. *Battles and Leaders* is still astonishingly good, despite its great age. Parish is an excellent modern social and political history.

3 For the experience of one group of improvised French soldiers in 1814 see my *Book of Sandhurst Wargames* (Hutchinson, London, 1982), chapter 3, 'Craonne'.

4 Michael Adams is an excellent guide to the morale fluctuations in the Eastern theatre. Lord's *They Fought For The Union*, pp. 217–28 gives a sound overview of morale in East and West. For General Sherman's comments on the fragility of the Bull Run armies, see Eisenschiml and Newman, 1, p. 63.

5 Jones, p. 28.

6 Watkins, p. 26. After First Manassas the ratio had been put at one Confederate being worth only five Federals (Wheeler, p. 48).

7 Michael Adams, p. 64.

8 Carl Davis, pp. 38–9.

9 See Batten for Napoleon III, Harsh for McClellan.

10 The eternal military debate between firepower and mobility was rarely articulated during the Civil War. Today, of course, it is of particular relevance to NATO's central front as the introduction of a new generation of high-technology firepower weapons is debated.

11 Lee reflected that 'No one knows how *brittle* an army is' (quoted in Hattaway and Jones, p. 308), although he might also have seen that Burnside's army was more brittle still.

12 We often forget that many Confederates came away from Gettysburg unaware that they had lost the battle. The moment when they had been open to a destructive counter-attack had been a very fleeting one (Downey, p. 167–8).

13 Relative costs of road, rail and water transportation are well discussed in Moore. See also the calculations in Sherman, 2, p. 389ff.

14 De Forest, p. 156.

15 See, e.g., Michael Adams, p. 34.

16 Wiley, *Billy Yank*, p. 323.

17 Blackford, p. 224.

18 Bull, p. 99. In McMurry, p. 100, however, we read that the reverse was the case, and the Western Confederates were better drilled than the Eastern.

19 For example, Bull, p. 191.

20 Listed in Dyer, 2, pp. 582ff.

21 Nisbet, p. 14. The Confederate Major Glover was killed at Winchester in 1864 during his 107th engagement. For the fragility of over-exposed veterans see, e.g., Douglas, p. 112.

22 Moseley, p. 349, takes it as evidence of high morale, but to the present author it seems based on a more pessimistic expectation. The latter view was shared by Charles Davis (p. 285) at Mine Run, and there is evidence of Confederate 'paper dog tags' from Jackson's Valley campaign which also places them in the context of pre-combat apprehension (Blackford, p. 104).

CHAPTER 2: COMMAND AND CONTROL

1 Hattaway and Jones, pp. 18–19, see the flexibility of nineteenth-century armies as a reason for the supposed indecisiveness of battle, but this argument cuts two ways. If the attack retains flexibility it presumably has a good chance of winning decisively, just as much as a flexible defence has a good chance of stopping it.

2 Hattaway and Jones, *passim*, are especially strong on the higher organisation of the two

sides. See also Shannon, Goff, and Jones for particular aspects.

3   Johnson, p. 18.

4   Rogers is informative on staff work, pp. 127ff, as are Wagner, pp. 25ff and Francis Lord, *They Fought for the Union*, pp. 56ff. The Gettysburg figures are from John W. Busey and David G. Martin, *Regimental Strengths at Gettysburg* (Gateway, Baltimore, 1982), p. 17.

5   Sherman, 2, p. 402.

6   *Battles and Leaders*, 2, p. 747.

7   Robert E. Lee complained of this (quoted in Rogers, p. 135), as had Halleck (p. 238) on the Union side.

8   Chamberlain, p. 98.

9   ibid., pp. 131–6, 154.

10   ibid., p. 128.

11   ibid., p. 148. The main calls are listed in Hardee, 1, pp. 221ff.

12   See Chamberlain, p. 130, for battlefield recognition of flags; Coggins, p. 11 and endpapers, and Todd, *passim*, for the heraldry itself.

13   For Lee's problems in 1862 see *Battles and Leaders*, 2, p. 361n., and Nisbet, p. 79; for Grant's in 1864 see Hattaway and Jones, p. 560. Francis Lord gives essential details of the topographical departments in his *Encyclopaedia*, pp. 95–9 – cf. Rogers, pp. 83–4.

14   Douglas, pp. 42–6, describes this process in his epic ride of 105 miles in 20 hours. English, *Confederate Field Communications*, pp. 44ff, shows how easy it was to get lost. Starr, 2, p. 463, for a 'man on a mule'.

15   *Battles and Leaders*, 22, p. 747.

16   Examples may be found in Pullen, p. 221, and Nisbet, pp. 46, 52, 212.

17   Sherman, 2, p. 397. Compare Hooker near Atlanta, who was rather less cautious (Bull, p. 156), although overall Union generals seem to have been much more cautious than Confederates (G. W. Redway, *The War of Secession*, Swan Sonnenschein, London, 1910, p. 145).

18   Quoted in Hattaway and Jones, p. 559.

19   The *Marschfeld* at Wagram, 1809.

20   Wilson gives a particularly vivid evocation of many of the central figures in the war.

21   A striking example of this was at Peach Tree Creek – Bull, p. 147.

22   Halleck, pp. 342ff.

23   ibid., pp. 118ff, and the discussion in Hagerman.

24   Boies, p. 84.

25   See the discussion of Jomini in Halleck, p. 124.

26   Wilson, p. 306.

27   Clausewitz, *On War*, ed. M. Howard and P. Paret (Princeton University, 1976).

28   Gustavus Smith, p. 348.

29   ibid., pp. 47–53.

30   ibid., p. 49.

31   Examples include 22nd Massachusetts at Gaines's Mill and Malvern Hill (Fratt, pp. 9–11), 14th Indiana at Kernstown (Baxter, p. 77), 12th Connecticut on the Opequon (De Forest, p. 182) and Longstreet's command at Spotsylvania (Sorrel, p. 240).

32   There is a very full description of this process in Jamieson, pp. 88–99.

33   For example, Watkins, p. 163, and Bull, pp. 149–50.

34   Naisawald, p. 266.

35   Pullen, pp. 162–3; Jamieson, pp. 104–5.

36   *Battles and Leaders* 4, pp. 545–67; *Thirty-Fifth Massachusetts*, pp. 265ff; Haskell pp. 73–80.

37   One is reminded of the initial US reluctance to deploy armour to Vietnam in the 1960s – Don Starry, 'Mounted Combat in Vietnam', in *Vietnam Studies* (Department of the Army, Washington DC, 1978).

38   For example, Seymour, p. 26.
39   For example, Sorrel, pp. 156 and 176 and Brown, *passim*.
40   Harsh, p. 69; Fishel, pp. 86, 105.
41   The subject of intelligence and signals may be studied in Fishel, Thompson and English, *Confederate Field Communications*. Rogers, and Lord, *They Fought For The Union*, also have useful information.
42   Fishel, p. 93.
43   See Hamlin, and Jennings Wise, p. 472. A similar failure to relay information at Shiloh is reported in Eisenschiml and Newman, 1, p. 173.
44   The cost of binoculars was not excessive – between $15 and $30 (Edwards, p. 90).
45   Thompson, p. 230; English, op. cit. pp. 17, 93, 133.
46   ibid., pp. 34, 65; Edwards, p. 226.
47   Thompson, pp. 231–5.
48   Grant, pp. 459–60.
49   English, op. cit., p. 136.
50   Sherman, 2, p. 398.
51   English, op. cit., pp. 78, 109.

CHAPTER 3: THE RIFLE

1   Out of 124,717 long arms retained by Union soldiers upon demobilisation in 1865, 7,424 (6 per cent) were breechloaders, 1,173 (1 per cent) were substandard rifles and muskets, 96,238 (77 per cent) were Springfield rifle muskets and 19,882 (16 per cent) were Enfields (Lord, *They Fought For The Union*, p. 166). Note that these figures do not include carbines.
2   Weller gives excellent details of range tests, by 'average' shots, with a wide variety of Civil War long arms.
3   Note Heth's 1858 target-training system, borrowed from the French *chasseurs* and reissued 1862 (Moseley, p. 297); also Willard's system, 1862 (Lord, *They Fought For The Union*, p. 27). Rudimentary target training had already been outlined in Scott's *Tactics*, e.g. 2, p. 211.
4   I find Richard Munday's analysis entirely convincing in this respect.
5   Exaggerated praise for snipers comes in, e.g., Shannon, 1, p. 143; a more balanced discussion is in Edwards, pp. 216–24. Sniper anecdotes occur in, e.g., Nisbet, p. 153, Bull, p. 155, Watkins, pp. 135, 143, Green, p. 127.
6   Bugeaud was only the best known of many French writers who advocated buck and ball – or even double-barrelled smoothbores – for greater firepower.
7   Lewis, p. 93; Carl Davis, p. 107.
8   ibid., p. 135.
9   ibid., pp. 65, 78; Lewis, pp. 64–8.
10   Thomas, Appendix 1. 1st Minnesota's armament is cited on p. 60.
11   New York regiments in 1861–2 had been armed as follows:

Total regiments	149	
Regiments with US 1842 muskets	31	21% (not known what proportion of these had been rifled)
Other smoothbores	13	9%
Other 2nd-rate rifles	34	22%
Enfield/Springfield rifles	71	48%
Breechloaders	0	

(Source: Lord, *They Fought For The Union*, p. 141)

12   Fremont's Western Army, 1861, had been especially badly armed (Edwards, pp. 133, 141).
13   Source: Thomas, Appendix 1.
14   Quoted in Commager, p. 92. Compare the extensive analysis of Confederate weapon availability in Fuller and Steuart, pp. 1–19, 104, 113.
15   Grant, p. 337.
16   Source: Fuller and Steuart, pp. 314–5.
17   Baxter, pp. 39, 105.
18   Charles Davis, pp. xviii, xxix, 16. The 'Winsor' was another name for the US 1841 Mississippi rifle, after its Windsor, Vermont, manufactory.
19   *Thirty-fifth Massachusetts*, p. 10.
20   ibid., p. 25.
21   McMurry, p. 64. Compare McKim, p. 135. Before Gettysburg Steuart's Confederate brigade included 1,941 soldiers present for duty with but 1,069 bayonets and 1,480 Springfield rifles. The balance was doubtless mostly accounted for by officers, musicians, etc., but it still surely represents a shortfall from total armament. Each armed man carried an average of 34 rounds, with as many more in the ordnance train.
22   Edwards, p. 342. After Chancellorsville the Confederates salvaged 12,000 arms (Jones, pp. 208, 213).
23   Green, p. 27.
24   Worsham, pp. 60, 165.
25   The salvaged arms from Chancellorsville, apparently, were allowed to rust by the Confederates (Fremantle, quoted in Edwards, p. 339). Note, however, that individual soldiers on campaign could also easily let their guns rust (Coe, p. 149).
26   Sources: Fuller and Steuart; Edwards; Carl Davis; Lewis, pp. 64–9, 227; Goff, pp. 14, 30, 62, 150; Albaugh, Benet and Simmons; Moseley, pp. 110–11.
27   Carl Davis, *passim*, explains the importance of these considerations, although he attributes too great a tactical superiority to the rifle musket over the smoothbore.
28   Source: Edwards, *passim*. Note that Carl Davis, p. 8, states that even rifle muskets were considered excessively expensive at the time. The costs in the South, too, were often much higher than in the North, as a result of the blockade. For example, revolvers sometimes changed hands for $500 (Fuller and Steuart, p. 237).
29   Lewis, p. 200.
30   Charles Davis, p. 324; Bull, p. 128.
31   Hattaway and Jones, p. 287; *Thirty-fifth Massachusetts*, p. 209.
32   Worsham, p. 17.
33   For example, McMurry, p. 141; Watkins, p. 114.
34   Thomas, p. 73, fn. 51. Against this we might cite the Union 14th Indiana, which was told in stirring tones that 'The South must become a wastefield' – and then sent into the Antietam holocaust with only ten rounds per man (Baxter, pp. 91, 99).
35   Bull, p. 36.
36   Gustavus Smith pp. 78ff.
37   For example, Confederate apologias for their initial defeat at Belmont (Pollard, 1, page not numbered officially – perhaps 190) – Beltzhoover has recourse to the bayonet! *Battles and Leaders* 1, pp. 349ff, tells a rather different tale.
38   In theory it was possible to fire three rounds per minute (Hardee, 1, p. 101), although in practice there were many factors to delay this – e.g. the reluctance of the second rank to fire through the first, and the fouling which set in after the first few rounds. I have been told that in modern target competitions it is rare for muzzle-loaders to achieve thirteen rounds in thirty minutes.
39   Pullen, p. 123; Baxter, p. 64.
40   *Battles and Leaders*, 2, p. 508.
41   Gustavus Smith, p. 109.

42 Bull, p. 150.
43 Watkins, p. 159.
44 Wiley, *Billy Yank*, pp. 82–3; Thomas, p. 46, states that 5th Alabama's sharpshooters fired 200 rounds each in the same battle.
45 *Battles and Leaders*, 4, pp. 173–4.
46 Gorgas, quoted in Moseley, p. 162.
47 Thomas, pp. 12–13.
48 The Napoleonic spectrum ranges from 8.7 rounds per hit (Maida, 1806, quoted in Holmes, p. 168) to 459 (Vitoria, 1813, quoted in Strachan, p. 54) – or even more (Busk pp. 18ff). The figure of about 20 rounds per hit quoted in B. P. Hughes, *Firepower* (London, 1974), is surely far too low as a general rule for battle results, although it is high for range practice.

For the Civil War, Shannon, 1, pp. 139–40, estimated that 300 lb of lead were needed for each hit, which works out at about 2,520 rounds per hit. This seems high, even for Sherman's stand-off skirmishers in 1864, and probably includes numerous rounds which were issued but not actually fired in combat. The Gettysburg figures, on the other hand, probably represent rather a low average for Civil War battles, due to the density of targets.

My own best guess is that in both Napoleonic times and in the 1860s the real average figure was somewhere between 100 and 1,000 rounds per hit, and that we cannot be much more precise than that. This finding does at least confirm the findings passed to the Richmond War Department in 1861 by unnamed French statisticians who showed that the practical lethality of smoothbore and rifle muskets was identical (Jones, p. 43).
49 McIntyre, p. 171.
50 Pullen, p. 28.
51 Source: Lewis, p. 85; cf. Hardee, 1, p. 36.
52 Bull, p. 149.
53 *Battles and Leaders*, 4, p. 173. Other examples are in De Forest, p. 122; Sears, p. 187.
54 Charles Davis, pp. 372, 374.
55 Cited in Shannon, 1, p. 137, from an article in the *US Service Magazine*, January 1865, p. 69. Moseley, p. 300, notes that an Army of the Potomac circular of 19 April 1864 made allusion to the same facts.
56 A whole pint of whiskey, however, may be smuggled into camp in a musket held vertically on the shoulder – anecdote quoted in Moseley, p. 119.
57 Michael Adams, pp. 23–31, 44–8. We may agree entirely with his argument without necessarily agreeing that the casualties at First Manassas are a relevant index of marksmanship between North and South.
58 See, e.g., Wiley, *Billy Yank*, pp. 27, 50; Shannon, 1, p. 173.
59 It is unnecessary to cite examples of snowball fights since, apart from revivalist prayer meetings, they seem to have been the most popular of all off-duty activities in the two armies – especially the Southern. Almost every personal memoir is full of them. Compare the British use of pine-cone fights in the aftermath of New Orleans, 1815.
60 Donald Smith, p. 25. Ward, p. 91, had a similar experience, and see Lord, *They Fought For The Union*, p. 35, for further examples.
61 Donald Smith, pp. 84, 179.
62 *Thirty-fifth Massachusetts*, p. 108. We are tempted to recall that at least one eminent cadet at Sandhurst in the 1950s successfully completed his two-year course without ever having fired his rifle.
63 Charles Davis, pp. 303, 95.
64 The point is well made by Coggins's sketches on pp. 24, 38–9.
65 McIntyre, p. 77. Compare Union admiration for slow and deliberate Confederate firing ('about one shot to our three hundred') quoted in Lord, *Encyclopaedia*, p. 15.
66 Wheeler, p. 36.
67 Poe and Seymour, p. 18.

CHAPTER 4: DRILL

1   Hardee, 2, p. 21; Zimmermann, pp. 3–11; Lord, *They Fought For The Union*, pp. 59ff. Bands were transferred from regiments to brigade HQs half-way through the war, although Donald Smith, pp. 48n. and 201, shows that this was delayed by eighteen months in 24th Michigan.
2   Nisbet, p. 135.
3   Zimmermann, p. 19. It was not unknown for parts of infantry regiments to become mounted for a spell, especially in the West – see, e.g. Anders, pp. 120–81.
4   Baxter, p. 37.
5   De Forest, p. 35. 'Nostalgia' was a well known military disease of the era, especially among boys who had never left their farms and families before joining the army – see Turbiana, *passim*.
6   Watkins, p. 25.
7   Hough, p. 81.
8   Livermore, p. 68; Lord, *They Fought For The Union*, p. 268; West, p. 38.
9   Charles Davis, p. 338; Pullen, p. 163; George Wise, p. 115. When 18th Missouri wanted to reduce to six companies from ten it encountered legal obstacles and had to make it clear it was only a 'temporary' arrangement – Anders, p. 78.
10   Wagner, p. 4. In 1814 Napoleon had made a virtue out of necessity, in this respect, in a similar manner.
11   *Thirty-fifth Massachusetts*, pp. 104ff.
12   Donald Smith's story of 24th Michigan is instructive. This regiment was accepted only grudgingly into the Iron Brigade (pp. 25, 38), but itself reacted violently when 167th Pennsylvania was later added to make good Gettysburg casualties (p. 157). There were also protests when the army corps it belonged to was merged with another (p. 176), although later drafts during the Petersburg siege were accepted with better grace (pp. 229, 230). The regiment ended the war by asserting its 'veteran' status in a vindictive manner when placed on guard over new drafts and bounty-jumpers (pp. 242, 244).
13   Charles Davis, pp. 228, 263 and appendix (nominal roll of recruits).
14   Sherman, 2, p. 388, complained of this, as did De Forest, p. 36. Many subsequent writers have repeated it.
15   Anders, pp. 120, 205, 291; Pullen, pp. 76, 152, 174, 236.
16   Sources: Fox, Dyer, Tancig, Livermore, *passim*, Parish, p. 132; and Shannon, 2, p. 227. The note of caution about these dated sources in Parish, p. 702n., can only be repeated here.
17   Livermore, pp. 68–70 (note that this applies to commissioned officers only). On the other hand we can find at least one occasion on which a Union charge was made entirely by officers and NCOs (Charles Davis, p. 338), while complaints that officers were scarce were continuous from Lee's headquarters (see, e.g., Sorrel, p. 279; McMurry, p. 71).
18   Among many complaints against elections see Nisbet, p. 131; McMurry, p. 32; Starr, *Union Cavalry*, vol 1, p. 223; Shannon, 1, p. 166ff.
19   Hattaway and Jones, p. 10. Robert K. Krick shows (in *Lee's Colonels*, Morningside Press, Dayton, Ohio, 1979, pp. xiii, xvii) how Military Academy graduates tended to reach the higher ranks – despite the devastations of the election system.
20   Michael Adams, pp. 97, 185.
21   Eisenschiml and Newman, 1, p. 63.
22   Shannon, 1, pp. 167–9.
23   Wiley, *Billy Yank*, pp. 321, 323.
24   *Thirty-fifth Massachusetts*, p. 220.
25   Quoted in Wagner, p. 40.
26   De Forest, p. 23.
27   Drill books are extensively listed in Lord, *They Fought For The Union*, pp. 44–52, and are well analysed in Moseley, Jamieson, and McWhiney and Jamieson. For the French

1831 revisions and subsequent debate, see my thesis on the French Army, pp. 141–73.
28   Scott, 1, pp. 6, 10, 69, 57.
29   ibid., p. 55.
30   ibid., 2, pp. 188, 217. *Grandes bandes* of skirmishers had been well known in the French Revolutionary wars, and their congruence with the 'spirit of the times' had been much appreciated in von Yorck's Prussian and Moore's British light infantry. For an American reflection of this movement, see William Duane's *Military Dictionary* (privately published, Philadelphia, 1810) and *A Handbook for Infantry* (ditto 1813).
31   By the end of the war the 'Gibraltar Brigade' included ten regiments (Baxter, p. 166) – more than Scott had thought made a whole corps!
32   Jamieson, pp. 30, 38–43, for the origins of Hardee's manual in the USA, to which we may add the analysis of Hardee's aims in Fratt, and Moseley's suspicion, p. 261, that it was Lieutenant S. V. Benét who actually wrote the manual – or rather translated it – and not Hardee at all.
33   Hardee, 1, p. 87.
34   Jamieson, p. 290.
35   See the whole Ellsworth document, and the history of its author's demise in Hattaway and Jones, p. 36. cf. Manning for the repercussions.
36   Jamieson, p. 119, and discussion in Chapter 6, below. At First Bull Run the 1st Maryland (CS) infantry jogged six miles to the battlefield, but on arrival 'all of us were nearly spent with the heat and the dust and the killing pace' – McKim, p. 35. See also Barron, p. 109, for 3rd Texas cavalry's very similar experience at Iuka.
37   Manning, pp. 10, 31; Lord, *They Fought For The Union*, pp. 137–8, and see also the opinion he quotes that French bayonet drill made men jump about 'like so many animated frogs' – *Encyclopaedia*, p. 55.
38   Manning, pp. 42–4. Admittedly, a number of Zouave regiments did retain their identity at the front, although this was not always in the tactical sense so much as in their distinctive dress and lavish supporting services. For example, Birney's Pennsylvania Zouaves had no less than 68 bandsmen and *vivandières* to minister to their spiritual needs (Lord, *They Fought For The Union*, p. 61). For the bloodthirsty and lawless character of the Wilson Zouaves, see Shannon, 1, p. 41.
39   There were only 19 regiments of white US regular infantry, few of which ever achieved their theoretical strength of two battalions (Fox, pp. 520–3), and Confederate regulars were considerably less numerous (Weinert, *passim*). One widespread, but scarcely encouraging, alternative method of determining élite status was by counting the casualties suffered in battle. The argument behind Fox's work is that a 'fighting regiment' was any which had 130 men killed in combat.
40   Bull, pp. 13–16, for Casey's camp; Jamieson, pp. 45–9, discusses his contribution. At one time the camp was called 'the reserve Army Corps', but it never functioned as such (Lord, *They Fought For The Union*, p. 32).
41   Upton, pp. 1, 98–110.
42   ibid., pp. 20, 143, 297ff. The square is on p. 203.
43   ibid., p. 111.
44   Fratt, p. 5, discusses this question for Hardee. By 1871 American manuals were trying to go further in the direction of 'tactics' even than Upton (Jamieson, pp. 224ff).
45   Upton, pp. 219, 309 – although admittedly the art of commanding an army corps is given only five out of Upton's 2,166 paragraphs.
46   Boies, p. 10; see also Charles Davis, p. 82, and McWhiney, p. 166.
47   Wiley, *Billy Yank*, pp. 26, 52–5, for some of the Civil War exercises. For European difficulties, see Strachan, pp. 18, 51; J. Houlding, *Fit For Service* (OUP, London, 1981), *passim*; and my thesis on the French army, pp. 220–2.
48   Malone, pp. 33 and 36, shows only three days' drill in four months during the winter 1861–2, but then the regiment was ordered 'to drill twist every day hear after' (p. 46).

49 McIntyre, p. 137.
50 Bull, pp. 133, 45. His figures tally with those of Sorrel, p. 242.
51 Bull, p. 151.
52 Worsham, p. 117.
53 *Thirty-fifth Massachusetts*, p. 46.
54 Poe and Seymour, p. 82.
55 Eisenschiml and Newman, 1, p. 333.
56 Bull, p. 50.
57 Commager, p. 372.
58 Charles Davis, p. 135.
59 Donald Smith, pp. 60-4. For a comparable performance in D Battery, 1st Rhode Island Artillery, at Second Manassas see Naisawald, p. 159.
60 De Forest, p. 60.
61 Eisenschiml and Newman, 1, p. 114.
62 Commager, p. 367.
63 Chamberlain, p. 20.
64 Nisbet, p. 22.
65 Wheeler, p. 92; Nisbet, p. 106; Watkins, p. 49; Boies, p. 32.
66 John Drewienkiewickz's lecture on 'The Effect of Breakfast' in the Wilderness battles of 1864 (delivered to the third Knuston Hall conference on military history) highlights the importance of timely food and efficient staff work.
67 Quoted in Moseley, p. 341. The original question was posed in 1885.
68 Sherman, 2, p. 394.
69 *Battles and Leaders*, 2, pp. 660, 662.
70 Eisenschiml and Newman, 1, p. 184.
71 Admittedly a few cases of sustained regular volley fire may be quoted, e.g. Arnold's from Turner's Gap; A Coy 17th North Carolina at Bentonville (Moseley, p. 383, and some other examples, pp. 383-5); 22nd Massachusetts at Fredericksburg (*Thirty-fifth Massachusetts*, p. 88). Even so, there is a problem of terminology here, since a prolonged fire at will is sometimes described in the sources as a single volley. Gustavus Smith, p. 51, cites a 'volley' which lasted fifteen minutes!
72 Wiley, *Billy Yank*, pp. 70-1; Bull, p. 56; Wheeler, p. 231; De Forest, pp. 111-12; Fratt, p. 10.
73 Watkins, p. 108; George Wise, p. 80; Worsham, p. 153; McMurry, p. 54.
74 Nisbet, pp. 177-82.
75 *Battles and Leaders*, 4, p. 163.
76 Anders, p. 109.
77 Donald Smith, p. 37, quoting the ubiquitous Dawes of 6th Wisconsin.
78 For example Wiley, *Johnny Reb*, p. 31.
79 Watkins, p. 34.
80 De Forest, p. 182n.
81 Examples of criticisms are in Moseley, pp. 336-40; Jamieson, pp. 183-4; Anders, pp. 206, 325; Pullen, pp. 34-5.

CHAPTER 5: THE BATTLEFIELD AND ITS FORTIFICATIONS

1 Jennings Wise, pp. 263, 268, 295.
2 Spalding, *passim*.
3 Sorrel, p. 237.
4 Humphreys, p. 55.
5 Ibid., p. 44.
6 Donald Smith, p. 75.

7 Chamberlain, p. 40.
8 Hattaway and Jones, p. 151, give a graph of soil saturation.
9 Sorrel, p. 176.
10 Worsham, p. 139.
11 Carter, Samuel III, *The Siege of Atlanta* (Bonanza, New York, 1973), p. 185.
12 Naisawald, p. 490.
13 Michael Adams, p. 77.
14 McIntyre, p. 76.
15 Sherman, 2, p. 394.
16 Bull, pp. 227, 239.
17 Haskell, p. 19. The 'Cuban war' in question was that of 1898 – but we marvel at the 'almost'.
18 Moore, *passim*, and Hattaway and Jones, p. 202.
19 Halleck had much to say in praise of fortifications (e.g. pp. 42, 77–84, 130, 148), especially for American militia troops. Jamieson, p. 77, makes an opposite point by asserting that fieldworks had not been used extensively in earlier American wars. He points to the mobile, open warfare of Mexico (1846) as the 'typical' way of proceeding. Perhaps we can reconcile the two by suggesting that open warfare represented the ideal of regular officers *if* they were confident in their troops; but fieldworks were an expedient frequently forced upon the others. Something close to this reading is favoured by Hattaway and Jones, p. 21n.
20 For Mahan see Hagerman, pp. 35–9, 45–8; Hattaway and Jones, pp. 11–14; Jamieson, pp. 28–9.
21 See my article 'The Strategic Challenge to the French Engineers 1815–51', in *Fort*, no. 5, spring 1978, p. 31. Compare an engineer's understanding of 'strategy' in Halleck, p. 62.
22 Hattaway and Jones, pp. 13–14.
23 Mahan's *actual* phrasing was that the 'spade, implementing the terrain, went hand in hand with rifle and bayonet' – quoted in Hattaway and Jones, p. 12.
24 The prevarications of the British caused a major shift in morale in favour of the Americans (Casey, p. 67), although American indiscipline remained a cause of great concern to their commander (Reilly, p. 310). Finally the eventual attack, coming after a fortnight's delay, was badly handled in several vital respects – for example, the assault troops were ordered to halt at the critical moment (Reilly, p. 299). Authorities seem agreed that the main damage to the attackers was done by artillery rather than small arms (Michael Adams, p. 42; Reilly, p. 307).
25 Michael Adams, pp. 78–9; Halleck, pp. 125, 141–8.
26 Halleck, p. 149.
27 Quoted in Moseley, p. 371, apparently from Livermore's *Northern Volunteers*, p. 922.
28 Descriptions of fieldworks may be found in, e.g., Nisbet, pp. 201–2; Wagner, p. 48; Charles Davis, p. 335; Bull, pp. 43–4; Patrick, p. 191.
29 Worsham, p. 133; Green, p. 137.
30 Worsham, p. 163; Coe, p. 161.
31 For example, the red soil fort, sixteen feet high from the bottom of the ditch, built at Cold Harbor in 1864 – *Thirty-fifth Massachusetts*, pp. 246–8.
32 Eisenschiml and Newman, 1, p. 582, show that the Confederate defences at Cold Harbor could be stepped over quite easily.
33 Green, pp. 108, 111, 151, 158.
34 Nisbet, p. 194; Sherman, 2, p. 288.
35 Chandler, appendix 1 p. 1118, suggests an average density of 6.7 men per yard for Napoleon's battles. Mahon, p. 61, quotes Maud's assumption that Civil War battlefields had one-eighth the density of Frederick the Great's.
36 See, e.g., Anders, p. 217, for 18th Missouri near Kenesaw. Sherman, 2, p. 55, stated

that the cleared area should be 'a hundred yards or more'.
37 A few artillery batteries had 'French rangefinders', on a private basis (Downey, p. 104). Admittedly, the evidence for the general lack of range markers is, of necessity, negative – we conclude they were not used because they are not mentioned in the literature. None the less it seems to be quite *deafeningly* negative.
38 Halleck, p. 65.
39 Quoted in Jamieson, p. 80.
40 Baxter, p. 143.
41 Hattaway and Jones, p. 87.
42 Sorrel, p. 128.
43 Hough, p. 204.
44 Coe, p. 160.
45 Commager, p. 942.
46 Anders, p. 226.
47 ibid., p. 229.
48 *Thirty-fifth Massachusetts*, pp. 232-3.
49 ibid., p. 250.
50 Jamieson, pp. 122-4.
51 ibid., p. 123.
52 Ashworth discusses all these problems at length, albeit in relation to the First World War.
53 Bull, p. 45.
54 Wagner, p. 93.
55 For example, Commager, p. 943; Fratt, p. 18; Donald Smith, p. 155.
56 Worsham, p. 33.
57 Chamberlain, p. 374. See also Anders, p. 325.
58 Quoted in Hattaway and Jones, p. 173.
59 Hood quoted in Jamieson, p. 134; Sheridan in Chamberlain, p. 161; Sherman, 2, pp. 396-7.
60 Hattaway and Jones, p. 581, mention that for a fortnight near Marietta the Army of the Cumberland was expending 200,000 rounds per day in skirmishing. Since this army formed only 60 per cent of Sherman's total force, we may estimate that it hit only some 3,000 Confederates out of about 5,000 made casualty in that time. This makes about one hit for every 935 rounds fired.

CHAPTER 6: THE INFANTRY FIREFIGHT

1 For example, the Union defenders of Beaver Dam Creek, 1862 – *Battles and Leaders*, 2, p. 329.
2 For example, the Union attack at Cold Harbor (Jennings Wise, p. 815); at Atlanta (Wagner, p. 94 – two examples); at Spotsylvania (Donald Smith, p. 192); at Cassville (Coe, p. 141). These cases are all suspiciously grouped together in 'the year of demoralisation', and we are entitled to wonder just how energetically the attacks in question were actually pushed.
3 Donald Smith, p. 36.
4 Compare accounts in Eisenschiml and Newman, 1, pp. 333-7; Fratt, p. 20; *Thirty-fifth Massachusetts*, pp. 86-8; Baxter, p. 119.
5 Watkins, p. 159.
6 Anders, p. 226; McKim, pp. 186, 206.
7 Commager, p. 942.
8 *Battles and Leaders*, 2, p. 510.
9 Eisenschiml and Newman, 1, pp. 296-7.

10   ibid., p. 441.

11   Gustavus Smith, pp. 105–40.

12   ibid., p. 114.

13   ibid., p. 116.

14   Assuming twenty rounds fired per man, these casualty figures would give one hit to every 350 Federal or every 138 Confederate bullets fired.

15   For European practice see my *Forward Into Battle*, pp. 12–62. For Mexico, Jamieson, pp. 1–24.

16   Hence the paradox of 'bayonet attacks' made without bayonets fixed – e.g., Pullen, p. 119; Eisenschiml and Newman, 1, p. 441. Buechler, p. 137, uses a similar case to argue the uselessness of the bayonet, not pausing to consider that soldiers would surely feel it made more sense to lunge at an opponent with a bayonet fixed than without.

17   Eisenschiml and Newman, 1, pp. 263–4.

18   Green, p. 128.

19   ibid., p. 123. The Orphan Brigade had already favoured this tactic in earlier battles – ibid., pp. 30, 68, 159.

20   *Battles and Leaders*, 2, p. 363.

21   McMurry, p. 48.

22   ibid., p. 141; see also pp. 72, 99, 183. Note an amazingly similar insistence on trophies by another 'charging' commander, Ulysses S. Grant (Porter, p. 103).

23   De Forest pp. 62–71.

24   ibid., p. 70.

25   Chamberlain, p. 45.

26   Quoted in McWhiney, p. 267.

27   Sherman, 2, p. 395.

28   Quoted in Eisenschiml and Newman, 1, p. 236. See also George Wise, p. 99, for another account of the same action.

29   Hattaway and Jones, p. 235, for a fascinating discussion of Lee's use of different phrases to describe his military aims, which normally fell short of annihilating the enemy. For a (rather less concentrated) analysis of Grant's aims, ibid., pp. 526–7.

30   Carl Davis, p. xi, specifically denies that the Civil War battles were decided at short range but he does not, alas, support this claim with evidence. Other authorities seem to favour quite close ranges – for example, Henderson said that most battles were fought under 100 yards (quoted in Moseley, p. 182) and Mahon, p. 57, claimed that attacks were stopped at 200–250 yards.

31   Strachan, p. 32.

32   My sources for these figures are all the books mentioned in the bibliography, wherever they give a clear statement of range by a participant in a Civil War battle. More systematic research still needs to be done on this – and many other tactical questions – by, for example, a close analysis of *OR*. Systematic research into these questions is also needed for the Napoleonic era, since much still remains obscure for that period of military history.

33   This point was ably made in Richard Munday's lecture, given in July 1985 at the Imperial War Museum for the Historical Breechloading Smallarms Association. See also my *Forward Into Battle*, pp. 137–43 for a theoretical commentary on the search for long-range weapons.

34   Worsham, p. 170.

35   *Thirty-fifth Massachusetts*, p. 169.

36   Hoffbauer, *passim*.

37   Very close-range fire could sometimes be sustained for long periods in the Civil War – for example, in the Devil's Den at Gettysburg we hear of 'an unflinching exchange of deadly fire' at fifty feet, i.e. 17 yards (Edwin C. Bennett, *Musket and Sword*, Coburn, Boston, 1900, p. 140. I am indebted to Steve Fratt for this reference). At Seven Pines 65th New York delivered a 'steady continuous accurate fire' at 25 yards (Gustavus Smith, p. 89);

At Spotsylvania 24th Michigan fired 5,000 rounds at fifty feet from the enemy (Donald Smith, pp. 197–8). Sometimes we even hear of the 'muzzles touching' those of the enemy (Eisenschiml and Newman, vol 1, p. 536).

38 Bull, p. 45, among others, describes the whistling of Minié balls.

39 Nisbet, p. 188; Pullen, p. 181.

40 Baxter, p. 98, for 14th Indiana at Bloody Lane, Antietam; De Forest, pp. 184–6, for 12th Connecticut on the Opequon.

41 Wagner, p. 112.

42 Formations are discussed in Moseley, pp. 357–69; Jamieson, pp. 89–100; Wagner, p. 90.

43 Jamieson, pp. 104–8; Wagner, pp. 92–3.

44 Halleck, p. 124.

45 A similar evolution during the First World War led commanders to call off attacks quickly in 1918, whereas in 1915–17 they had often allowed them to drag on for weeks after maximum results had been achieved (see Blaxland, G., *Amiens 1918*, London, 1968, p. 196).

46 Halleck, pp. 125, 148.

47 The use of this tactic at Fort Donelson is discussed in *Battles and Leaders*, 1, pp. 422–4, and in Wagner, pp. 87–8.

48 G. M. Trevelyan, *Garibaldi and the Thousand* (London, 1909) for the battle of Calatafimi; Hoffbauer, p. 102, for an early example of a 'Prussian rush'. Strachan, p. 27, mentions a 'Sepoy Rush' used against the Sikhs in the 1840s.

49 Wagner, p. 87; Mahon, p. 63 (following Henderson); Moseley, p. 356 (although contradicted by his own p. 369).

50 McMurry, p. 42.

51 De Forest, p. 190.

52 Worsham, p. 152.

53 Fratt, pp. 14–15.

54 Charles Davis, pp. 166–8. At Winchester this regiment actually had assaulted and carried a line of enemy trenches with skirmishers only – but the trenches were found to be unoccupied at the time (ibid., p. 30).

55 Quoted in Hattaway and Jones, p. 584. Note, however, that when a major assault *was* intended it was not uncommon for all skirmishers to be withdrawn completely – see, e.g., Conyngham, p. 75; Bull, p. 133.

56 Jamieson, p. 121; Mahon, p. 61.

57 Eisenschiml and Newman, 1, p. 114.

58 Nisbet, pp. 203–4.

59 ibid., p. 200. Compare a successful trench raid at Petersburg which included impersonation of the enemy (Conyngham, p. 248).

60 Green, p. 104. Haskell, pp. 73–4, said something similar about the vulnerability of the Confederate line at Petersburg.

61 Hancock was lucky with his mist, but did not realise it. He actually delayed an hour, hoping it would clear, and then had his men cheer loudly – warning the enemy – while they were still several hundred yards short of the objective – (Grant, p. 475).

62 *Thirty-fifth Massachusetts*, p. 83.

63 Bull, p. 157.

64 Wagner, p. 438.

65 Malone, p. 86. McKim, pp. 195–200, describes this action from the viewpoint of a neighbouring division, and it is clear that considerable success was achieved.

66 For example, twice in a night at Hazel Grove, Chancellorsville (Jennings Wise, pp. 488, 495); Wagner, p. 435 for Big Bethel.

67 Sorrel, p. 142.

68 Boies, p. 48; *Battles and Leaders*, 3, p. 690.

69 McKim, pp. 199–205.
70 Charles Davis, p. 109.
71 Wheeler, p. 235.
72 Charles Davis, p. 109.
73 Pullen, p. 185. On p. 191 we read of a whole army corps being yelled into retreat.
74 Jennings Wise, p. 225.
75 References are, respectively, Wiley, *Johnny Reb*, p. 361 (also includes good phonetic discussion of the Rebel yell itself); Baxter, p. 98; Barron, p. 46; McMurry, p. 63; Jamieson, p. 80; McKim, p. 36; Jamieson, p. 80.
76 Perhaps predictably, the careful 'flanker' Sherman disapproved of noisy assaults – quoted in Eisenschiml and Newman, Vol 1, p. 68. He was, however, an exception to the general rule.
77 Fratt, p. 12.
78 *Thirty-fifth Massachusetts*, p. 85.
79 ibid., pp. 88–9.
80 Worsham, p. 152.
81 *Thirty-fifth Massachusetts*, p. 58. They received a very random selection from Casey's holdings in infantry accoutrements.
82 ibid., p. 238; McIntyre, p. 55.
83 Wheeler, p. 236. A variant is the Confederate joking before Gettysburg – 'When we charge the intrenchments, boys, recollect the crackers inside' (McKim, p. 146).
84 Upton, p. 20.
85 ibid., p. 98.
86 ibid., p. 143.
87 De Forest, pp. 182–3.
88 Jamieson, p. 120; George Wise, pp. 67, 80.
89 Coe, p. 131.
90 Only Ardant du Picq, in France, made a truly penetrating study of the problems associated with *chasseur* tactics.

CHAPTER 7: ARTILLERY

1 This marvellous image was picked up bodily by Conrad in *The Heart of Darkness* (1902) and transferred to a French cruiser shelling an anonymous section of the African bush, *c.* 1890. It stands for the blindness and impersonality of long-range, high-tech firepower. Conrad's French cruiser was in turn picked up bodily by Francis Ford Coppola and turned into an Arclight strike in his film *Apocalypse Now* (1979) – which was all about the war in Vietnam.
2 Jennings Wise, p. 154; Naisawald, p. 24.
3 Naisawald, pp. 33, 332; Wise, p. 574; Downey, pp. 11, 250 and – for Confederate refusal to accept promotion out of the artillery – p. 45.
4 Downey, p. 124; see also Longacre's biography for Hunt.
5 Downey, p. 255.
6 ibid., p. 214. See also *Battles and Leaders*, 3, pp. 434–7, for the order of battle.
7 Downey pp. 110, 137, 215, 251.
8 Compare the order of battle at Appomattox – *Battles and Leaders*, 4, pp. 748–51.
9 Lauerma, *passim*, for the French learning process, and cf. my *French Artillery* for the Napoleonic follow-on. Against this, the Koskiusko translation appears to have given Americans but a pale reflection of European developments. By 1870, on the other hand, Prussian practice was quite systematic in emphasising artillery concentrations from the very start of any battle (Hoffbauer, *passim*).
10 Naisawald, p. 25.

11   ibid., p. 311.
12   Jamieson, pp. 148–52.
13   Wise, pp. 488–509; Naisawald, pp. 294–309.
14   Wise, pp. 295–312, 356–99; Naisawald, pp. 184–229, 236–66.
15   *Battles and Leaders*, 3, p. 364; Downey, pp. ix, 127–55, 212.
16   Naisawald, p. 536. Compare McMurry, p. 99, for Confederate agreement on this point.
17   The 'worst case' must have been that of Confederate gunners trying to hold the very steep and obstructed Missionary Ridge.
18   By 1864 we start to find a few experiments in indirect/blind firing – see, e.g., Haskell, p. 73; Naisawald, pp. 475, 477, 485–6, 497; Wise, p. 826.
19   For example Wise, pp. 218, 249, 263, 273, 303, 386, 395, 397, 490, 520–1, 615, 656, 760, 778, 799; Naisawald, pp. 59, 65, 72, 77, 94, 113, 130, 156, 173–5, 194, 224, 259, 294, 321.
20   Bull, p. 111; Wise, p. 536; Naisawald, pp. 8, 57, 96, 322.
21   Naisawald, pp. 35–6, 464, 489–90; Jamieson, p. 154.
22   Naisawald, pp. 73, 460; Coggins, p. 67.
23   Moseley, pp. 198–205.
24   Haskell, p. 63.
25   Jamieson, p. 139; Wise, p. 390. Naisawald, p. 266, puts the Fredericksburg figure at 20 per cent, not 50.
26   Wise, p. 268.
27   Downey, p. 261, quotes a Confederate gunner who estimated that artillery attacks on fortified troops were effective only up to 1,000 yards range.
28   The Union army on Mine Run in late 1863 left all its rifled cannon behind, taking only Napoleons for the battle (Naisawald, p. 460). For the disconcerting way canister threw up dirt into attackers' faces see Green, p. 95; A. F. Ford, *Life in the Confederate Army* (New York, 1905), pp. 58–9.
29   Coe, p. 152.
30   Gustavus Smith, p. 87.
31   Haskell, pp. 67–8.
32   We might today attribute more of Hancock's success to the mist, but at the time the Confederates' lack of artillery was given pride of place (Wise, p. 789).
33   Wagner, p. 113, advised against long-range infantry fire at artillery – as late as 1895.
34   Naisawald, pp. 53, 70, 73, 77.
35   ibid., p. 174.
36   Organisation is covered in Wise, pp. 156–7, 199, 241, 332–56, 821; and Naisawald, pp. 28–32, 148, 178–84, 229–32, 266, 326–32.
37   Naisawald, pp. 174, 166.
38   Wise, p. 132, says that forty men were hit; Naisawald, p. 15, says less than thirty-two.
39   Naisawald, pp. 198–211.
40   ibid., pp. 261, 311.
41   Haskell, p. 67; Downey, pp. 17, 21.
42   *Battles and Leaders*, 2, pp. 314–6.
43   Wise, pp. 273, 546.
44   Naisawald, p. 443; Wise, p. 690.
45   Naisawald, p. 266; Wise, pp. 390, 544.
46   Fox, p. 36; Poe and Seymour, p. 10. Compare general statistics from Livermore, *passim*, used in my Appendix II.
47   Fox, p. 125.
48   Wise, p. 268.
49   ibid., pp. 263, 268. In the case of Pickett's charge at Gettysburg the attackers did not stop but swerved away from the guns – Downey, p. 150.

50 For example, Edwards, pp. 217–9.
51 Naisawald, p. 456.
52 A favourite maxim of General Foy, cited in Lauerma.
53 Wise, p. 167.
54 Haskell, p. 82.
55 Wise, pp. 301, 383.
56 Naisawald, pp. 104, 198–223, 246–66.
57 *Battles and Leaders*, 3, p. 360.
58 Wise, pp. 221–31.
59 The infantry's interference was certainly resented mightily – see Naisawald, pp. 24, 60, 184, 328.
60 ibid., pp. 27, 665.
61 Wise, pp. 284, 301, 342. By a statistical quirk, Lee had no less than 73 guns per 1,000 men at the battle of Sheperdstown (ibid., p. 319).
62 Sherman, 2, p. 396; cf. *Battles and Leaders*, 4, pp. 472–3, 696–8, for the order of battle in the two final Western campaigns.

CHAPTER 8: CAVALRY

1 Miller, p. 66.
2 Starr, 1, pp. 317, 325.
3 ibid., p. 275; *Battles and Leaders*, 2, pp. 344–6, 364, 430.
4 Starr, *Union Cavalry*, 1, p. 295.
5 ibid., p. 359. See Jennings Wise, p. 488, for how little the Confederates noticed this attack.
6 Starr, *Union Cavalry*, 1, p. 440; *Battles and Leaders*, 3, pp. 394–5.
7 Starr, *Union Cavalry*, 1, p. 249.
8 Gray, pp. 5–8; Emmert, *passim*.
9 ibid., p. 34.
10 Starr, *Union Cavalry*, 1, p. 381; cf. Barron, p. 223, for ground-clearing at Lovejoy Station.
11 ibid., 1, pp. 235–7.
12 Michael Adams, p. 41–3, casts doubt on the stereotype of the 'Southern horseman', although it is accepted by Starr, *Union Cavalry*, 1, p. 211.
13 For example, Barron, p. 53. The point is well discussed in Starr, *Union Cavalry*, 1, pp. 222–33.
14 ibid., p. 216.
15 ibid., p. 273.
16 Hattaway and Jones, *passim*, is the best source for the Union doctrine of raiding. For the Chancellorsville detachment see Starr, *Union Cavalry*, 1, pp. 351–8.
17 Starr, *Union Cavalry*, 1, pp. 26–46; Miller, p. 18.
18 Starr, 'Cold Steel', pp. 116–22; for the superiority of revolvers over repeating carbines in a charge see Barron, p. 234.
19 For example, Hampton's Legion included six infantry, four cavalry companies and a battery. It was split up in 1862, according to arm – Zimmermann, pp. 11, 19.
20 Barron, pp. 252–3.
21 Jennings Wise, p. 596.
22 In the 1864 campaign in Virginia a Union infantry corps averaged some 20,000 men in four divisions, whereas Sheridan's cavalry corps had about 12,000 in three. We could estimate from this that an infantry corps might be able to place perhaps 18,000 men in its musketry firing line whereas the cavalry corps might manage only 8,000.
23 Chamberlain, p. 63, suggests that it was Grant who stopped Sheridan departing for a

raid, but Starr, *Union Cavalry*, 2, pp. 426, 432, gives convincing details to show it was the other way round.
24 Chamberlain, pp. 64, 90, 180.
25 Starr, *Union Cavalry*, 2, pp. 445–50.
26 ibid., pp. 459–65.
27 ibid., pp. 470–3.
28 McKim, p. 275; Stonebraker, p. 102. The unfortunate victim was the only man hit by the fire of 'thousands' of Spencers.
29 Starr, *Union Cavalry*, 2, pp. 477–85.
30 Edwards, p. 162.
31 Starr, *Union Cavalry*, 2, p. 467; cf. p. 463: 75 rounds was the normal load for each of the men, and they were reluctant to obey an order to carry 125 rounds.
32 ibid., p. 118.

BIBLIOGRAPHY

ADAMS, GEORGE W., *Doctors in Blue* (Schuman, New York, 1952). A medical history of the Union side, making full play with the returns for wounds in the Wilderness campaign.

ADAMS, MICHAEL C.C., *Our Masters the Rebels* (Harvard University Press, Cambridge, Mass., 1978). A convincing analysis of low morale in the Army of the Potomac.

ALBAUGH, W.A. III, BENET, H. jr., and SIMMONS, E.N., *Confederate Handguns* (Riling & Lentz, Philadelphia, 1963). Authoritative listing and discussion.

AMBROSE, STEPHEN E., 'Dennis Hart Mahan', in *Civil War Times Illustrated*, November 1963, pp. 30–35. A eulogy from one who believes that engineers should be allowed to shape the curricula of military academies.

ANDERS, LESLIE, *The Eighteenth Missouri* (Bobbs-Merrill, Indianapolis, Kansas City and New York, 1968). Shows the wilder Western experience, in telling contrast to the orderly East.

ARDANT DU PICQ, CHARLES J., *Battle Studies* (Translated by John N. Greely and Robert C. Cotton, Macmillan, New York, 1921). The most advanced '*chasseur* thinking' of the French Second Empire.

ARNOLD, JIM, 'Eyewitness Accounts of History', in *The Courier*, vol. 3, no. 6, May–June 1982, p. 31. Reprint of an action at Turner's Gap, September 1862, from 6th Wisconsin Volunteers (Memoirs of Major Dawes, 1890).

ASHWORTH, TONY, *Trench Warfare, 1914–18: The Live and Let Live System* (Macmillan, London, 1980). Excellent use of sociological method to show how tacit truces arise in combat.

BAILEY, DE WITT, articles in *Guns Review*, October 1969, p. 401; April 1970, p. 134; July 1970, p. 265; August 1970, p. 300. Accounts of the various British long arms employed in the Civil War.

BARRON, SAMUEL B., *The Lone Star Defenders* (Neale, New York, 1908; reprinted Zenger, Washington DC, 1983). Memoirs from the 3rd Texas Cavalry.

BARTON, MICHAEL, *Goodmen: The Character of Civil War Soldiers* (Pennsylvania State University Press, 1973). Statistical and linguistic analysis of Civil War writings.

BATTEN, SIMON, 'Napoleon III as Military Commander' (Unpublished undergraduate thesis, Oxford University, 1985). A wide-ranging discussion of Napoleon *le petit*'s contribution to the military art, especially in 1859.

Battles and Leaders: JOHNSON, ROBERT U. and BUEL, CLARENCE C. (eds), *Battles and Leaders of the Civil War* (The Century Co., New York, 4 vols, 1884 and many reprints). The indispensable guide to the subject, written by many of the 'leaders' themselves from both North and South.

BAXTER, NANCY, *Gallant Fourteenth* (Pioneer Study Center, Traverse City, Indiana, 1980). Good history of a true 'Western' regiment, 14th Indiana.

BERGER, C., *Broadsides and Bayonets* (Oxford University Press, New York, 1961). Useful analysis of the propaganda of the American Revolution.

BILLINGS, JOHN D., and McCARTHY, CARLTON, *Soldier Life in the Union and Confederate Armies* (first published separately in, respectively, Boston, 1888, and Richmond, 1882; collected and edited by Philip Van Doren Stern for Premier Civil War Classics, Fawcett, New York, 1961). The soldier's life on campaign, apart from the battles.

BLACK, ROBERT C. III, *The Railroads of the Confederacy* (University of North Carolina Press, 1952). A splendid merger of railroad history and military history.

BLACKFORD, SUSAN L., *Letters from Lee's Army* (Scribner's, New York, 1947; reissued by Perpetua, 1962). Experiences of an officer in 2nd Virginia Cavalry.

BOIES, ANDREW J., *Record of the 33rd Massachusetts Volunteer Infantry* (Fitchburg, 1880). One of the less revealing memoirs from Sherman's army.

BROWN, R. S., *Stringfellow of the Fourth* (Crown, New York, 1960). Uninformative and sensationalised account of the life of a spy.

BUECHLER, JOHN, '"Give 'em the Bayonet" - A Note on Civil War Mythology', in *Battles Lost and Won*, ed. John T. Hubbell (Greenwood Press, Westport, Conn., 1975), pp. 135–8. A superficial treatment, lacking in understanding.

BUGEAUD, T. R., *Aperçus sur quelques détails de la guerre* (24th edn, Paris, 1873). A seminal text for shock tactics, written in the 1820s.

BULL, RICE C., *Soldiering*, ed. K. Jack Bauer (Presidio Press, San Rafael, Calif., 1977). Diaries from 123rd New York.

BUSK, HANS, *The Rifle and How To Use It* (8th edn, Routledge, Warne & Routledge, London, 1861). One of the optimistic mid-nineteenth-century advocates of long-range marksmanship.

CASEY, POWELL A., *Louisiana in the War of 1812* (Privately published, Baton Rouge, La., 1963). Detailed, well researched local military history.

CASEY, SILAS, *Infantry Tactics for the Instruction, Exercise and Manoeuvres of the Soldier, a Company, Line of Skirmishers, Battalion, Brigade, or Corps d'Armée* (3 vols, Evans & Cogswell, Columbia, SC, 1864), Vol. 3. A Confederate edition of the 1862 Union revision of Hardee.

CHAMBERLAIN, JOSHUA L. *The Passing of the Armies* (Putnam, New York, 1915; new edn, Morningside Bookshop, 1974). A rambling rhetorical account of the US V Corps in the Appomattox campaign, 1865, which nevertheless contains many gems.

CHANDLER, DAVID G., *The Campaigns of Napoleon* (Macmillan, London, 1967). A classic modern reference work.

COE, HAMLIN A., *Mine Eyes Have Seen The Glory* (Associated University Presses, Cranbury, NJ, 1975). Memoirs of a sergeant in 19th Michigan.

COGGINS, JACK, *Arms and Equipment of the Civil War* (Doubleday, New York, 1962). Probably the classic popular source for 'hardware', greatly assisted by the author's own illustrations and his sound grasp of Civil War tactics. Marred by lack of scholarly apparatus and excessive dependence on standard reference works, e.g. Lord.

COMMAGER, HENRY S. (ed.), *The Blue and the Gray* (Bobbs-Merrill, Indianapolis and New York, 1950). Excellent collection of snippets from contemporaries. Not to be confused with the earlier Kirkland or Gerrish and Hutchinson collections with the same name.

CONYNGHAM, D. P., *The Irish Brigade and its Campaigns* (Cameron & Ferguson, Glasgow, n.d. (1866?)). Detailed unit histories from one part of the Union's 'Celtic heritage'.

CRANE, STEPHEN, *The Red Badge of Courage*, ed. S. Bradley, R. C. Beatty, E. H. Long (Norton Critical Edition, New York, 1962). *The* Civil War novel. The discussion of literary influences in this edition is fascinating.

DAVIS, CARL L., *Arming the Union - Small Arms in the Union Army* (Kennikat Press, Port Washington, NY, 1973). Scholarly modern discussion of the problems of procurement.

DAVIS, CHARLES E. jr, *Three Years in the Army* (Estes & Lauriat, Boston, Mass., 1894). A very thorough history of 13th Massachusetts Volunteers.

DINTER, ELMAR, *Hero or Coward* (Cass, London, 1985). A modern discussion of combat psychology.

DONALD, D. (ed.), *Why the North Won the Civil War* (University of Louisiana, Baton Rouge, 1960). A stimulating collection of essays by distinguished writers, mainly concerning the strategic and political background to the war.

DOUGLAS, H. KYD, *I Rode with Stonewall* (new edn, Putman, London, 1945). Interesting memoirs of a Confederate staff officer.

DOWNEY, FAIRFAX, *The Guns at Gettysburg* (McKay, New York, 1958). Lively, sometimes 'purple' prose to accompany Wise, J. and Naisawald on this subject.

DUPUY, T. N., *Numbers, Predictions and War* (Macdonald & Jane's, London, 1979). An egregious attempt to reduce battle to statistics.

DYER, F. H., *A Compendium of the War of the Rebellion* (new edn, 3 vols, Yoseloff, New York, 1953). A monumental assemblage of statistics for the Union armies, particularly useful for the thumbnail histories of each regiment.

EDWARDS, WILLIAM B., *Civil War Guns* (Stackpole, Harrisburg, Pa., 1962). Lively popular coverage, although somewhat disjointedly arranged.

EISENSCHIML, OTTO, and NEWMAN, RALPH, *Eyewitness – The Civil War* (2 vols, including Vol. 1, first published in 1947 as *The American Iliad*) (Grosset & Dunlap, New York, 1956). A magnificent collection of 'tactical snippets' from the Civil War battles.

ELLSWORTH, EPHRAIM E. *Manual of Arms for Light Infantry, Adapted to the Rifled Musket* (Chicago, 1859, arranged for the US Zouave Cadets). An insight into the thinking of this remarkable warrior.

EMMERT, H. D., 'Civil War Cavalry', in *Wargamer's Digest*, April, August and September 1980. Useful basic analysis.

ENGLISH, JOHN A., 'Confederate Field Communications' (unpublished MA thesis, Duke University, Durham, NC, 1964). Workmanlike coverage of a broad subject.

ENGLISH, JOHN A., *On Infantry* (Praeger, New York 1981). Infantry tactics from the 1860s to the present, concentrating especially on the Second World War.

FISHEL, EDWIN C., 'The Mythology of Civil War Intelligence', in *Battles Lost and Won*, ed. John T. Hubbell (Greenwood Press, Westport, Conn., 1975), pp. 83–105. Excellent overview of a vital subject.

FOREST, JOHN W. De, *A Volunteer's Adventures*, ed. J. H. Croushore (Yale University Press, 1946; reprinted Archon Books, New York, 1970). Written by a novelist, these memoirs of 12th Connecticut Volunteers are probably the nearest thing we have to 'faction' that is nevertheless hard fact. Few books in any era can say as much.

FOSTER GLEASON, F. W. 'Arms For the Men', in *Ordnance*, July–August 1961, pp. 62–5. A useful, if not outstandingly scholarly, orientation.

FOX, WILLIAM F., *Regimental Losses in the American Civil War* (Brandow, New York, 1898; reprinted Morningside Bookshop, 1974). Extensive statistics on deaths in the Union army, including the controversial '300 fighting regiments' which lost 130 or more killed in action.

FRASSANITO, WILLIAM A., *America's Bloodiest Day* (Mills & Boon, London, 1978). Brilliantly detailed and documented study of the haunting Antietam photographs.

FRATT, STEVE, 'American Civil War Tactics: The Theory of W. J. Hardee and the Experience of E. C. Bennett', in *Indiana Military History Journal*, vol. 10, no. 1, January 1985, pp. 4–17. A perceptive explanation of the value of drill on the battlefield, based on Bennett's *Musket and Sword* (Coburn, Boston, 1900) – memoirs from 5th, then 22nd Massachusetts Volunteers.

FULLER, C. E., and STEUART, R. D., *Firearms of the Confederacy* (Standard Publications, Huntington, WVa, 1944). Exceptionally well documented and perceptive. See pp. 113–30 for Gorgas's vital memoir.

GLUCKMAN, ARCADI, *United States Muskets, Rifles and Carbines* (Ullbrich, Buffalo, NY, 1948). Listing of models and contractors.

GOFF, RICHARD D., *Confederate Supply* (Duke University Press, Durham NC, 1969). Scholarly modern analysis of an intractable problem.

GRANT, ULYSSES S., *Personal Memoirs of U. S. Grant* (new edn, Webster, New York, 1894). Lucid exposition from a dying ex-President.

GRANT, U. S. III, 'The Military Strategy of the Civil War', in *Military Analysis of the*

Civil War (KTO Press, Millwood, NY, 1977), pp. 3–15. A useful, if somewhat dated, general analysis.

GRAY, ALONZO, *Cavalry Tactics as Illustrated by the War of the Rebellion* (Fort Leavenworth, Kansas, 1910). A pithy discussion of 'tactical snippets' from the Civil War, intended to give guidelines for the future.

GREEN, JOHNNY, *Johnny Green of the Orphan Brigade*, ed. A. D. Kirwan (University of Kentucky Press, 1956). Memoirs from a Kentucky outfit which believed in shock action.

GRIFFITH, PADDY, *Forward Into Battle* (Bird, Chichester, 1981). Reinterpretations in tactical history between 1808 and 1973.

GRIFFITH, PADDY, *French Artillery 1800–1815* (Almark, London, 1976). A short summary.

GRIFFITH, PADDY, 'Military Thought in the French Army 1815–51' (unpublished D Phil thesis, Oxford University, 1976). Discussion of a peacetime debate rather than of battlefield practice.

HAGERMAN, EDWARD, 'From Jomini to Dennis Hart Mahan: The Evolution of Trench Warfare and the American Civil War', in *Battles Lost and Won*, ed. John T. Hubbell (Greenwood Press, Westport, Conn., 1975), pp. 31–54. A seminal article. Unfortunately I have been unable to obtain sight of his doctoral thesis in the time available for my research ('The Evolution of Trench Warfare in the American Civil War', Duke University, 1965).

HALLECK, H. WAGER, *Elements of Military Art and Science* (Appleton, New York, 1846; reprinted Greenwood Press, Westport, Conn., 1971). Revealing distillation of the way French theory was perceived by a career engineer.

HAMLIN, A. C., *The Battle of Chancellorsville – Jackson's Attack* (Privately published, Bangor, Maine, 1896). The historian of XI Corps, Army of the Potomac, argues that his formation won the battle. It would be nice to believe him.

HARDEE, W. J., *Rifle and Light Infantry Tactics for the Exercise and Manoeuvres of Troops* (2 vols, Lippincott, Philadelphia, Pa, and Griffith, Nashville, Tenn., 1861 edn). The Civil War's translation of the most advanced French drill. (The first edition, of 1855, has been reprinted by the Greenwood Press, Westport, Conn., 1971).

HARSH, J. L., 'On The McClellan-Go-Round', in *Battles Lost and Won*, ed. John T. Hubbell (Greenwood Press, Westport, Conn., 1975), pp. 55–72. Useful revisionism on 'the new Napoleon'.

HASKELL, JOHN C., *The Haskell Memoirs*, ed. G. E. Govan and J. W. Livingood (Putnams, New York, 1960). Memoirs of a South Carolina artillery officer in Lee's artillery.

HATTAWAY, HERMAN, and JONES, ARCHER, *How the North Won* (University of Illinois Press, 1983). A splendid, if rather repetitive, interpretation of the operational and strategic levels of action.

HOFFBAUER, E., *The German Artillery in the Battles Near Metz*, translated by Captain Hollist, (King & Co, London, 1874). An excellent technical history of three key battles in 1870.

HOLMES, RICHARD, *Firing Line* (Cape, London 1985). 'Tactical snippets', mainly from the twentieth century, concerning the individual's perception of and reaction to combat. (Published in USA as *Acts of War*.)

HOUGH, ALFRED L., *Soldier in the West*, ed. R. G. Athearn (University of Pennsylvania Press, 1957). Memoirs from a rabidly anti-secesh officer who saw many campaigns but few battles.

HUMPHREYS, ANDREW A., *The Virginia Campaign of 1864 and 1865*, number 12, Campaigns of the Civil War series (new edn, Scribners, New York, 1907). Excellent staff history from Grant's high command.

JAMIESON, PERRY D., 'The Development of Civil War Tactics' (unpublished PhD thesis, Wayne State University, Detroit, 1979). An excellent and very carefully researched analysis which suffers only from the author's wilful persistence in the false theory that the rifle musket changed tactics.

JOHNSON, LUDWELL H., 'Civil War Military History – A Few Revisions in Need of Revising', in *Battles Lost and Won*, ed. John T. Hubbell (Greenwood Press, Westport, Conn., 1975), pp. 3–18. A stimulating discussion of strategy and historiography.

JOMINI, A. H., *Précis de l'art de la guerre* (New edn, Paris, 1855). Often claimed as an important source for Civil War ideas about generalship, but probably as little understood in the 1860s as in the 1980s.

JONES, JOHN B., *A Rebel War Clerk's Diary*, new edn, condensed by E. S. Miers (Sagamore Press, New York, 1866). Prejudiced ramblings from Richmond which have somehow become a classic.

KIRKPATRICK, D. L. I., 'Lanchester and Real Battles', in *Royal United Services Institute Journal*, vol. 130, no. 2, June 1985, and follow-up discussion in vol. 130, no. 4, December 1985. Yet another statistical model for Civil War tactics, showing that attrition rates depended on numerical strength in conformity with Lanchester's linear law, not his square law.

KOSKIUSKO, *Manoeuvres of Horse Artillery*, translated by Colonel J. Williams (US Military Philosophical Society, New York, 1808). An early discussion of artillery tactics which stresses the local difficulties in America.

LAUERMA, MATTI, *L'Artillerie de campagne Française pendant les guerres de la révolution* (Katara, Helsinki, 1956). Excellent 'gunner history' of the rise of concentrated batteries in the 1790s.

LEONARD, THOMAS C., *Above the Battle – War Making in America from Appomattox to Versailles* (Oxford University Press, New York, 1978). A well sustained attack by a liberal upon the reticences generated by war-making, although only the first thirty pages concern the Civil War. Is this itself a reticence?

LEWIS, BERKELEY R., *Small Arms and Ammunition in the US Service 1776–1865* (Smithsonian Institution, Washington DC, 1956; new edn, 1968). One of the best treatments of the 'hardware', especially ammunition.

LIEBER, FRANCIS, *Instructions for the Government of Armies of the United States in the Field* (US War Department/Van Nostrand, New York, 1863). 'The Laws of War' as perceived at the time.

LIVERMORE, THOMAS L., *Numbers and Losses in the Civil War in America 1861–5* (Massachusetts Military Historical Society, 1897; new edn, with introduction by Edward E. Barthell jr, Indiana University Press, Bloomington, 1957). A dedicated Federal tries to unravel the vital statistics of the war, with patchy success.

LOIR, CAPITAINE, *Cavalry – Technical Operations* (official translation, HMSO, London, 1916). Excellent discussions of cavalry in the Franco-Prussian War, proving the decisiveness of massed charges in 1870.

LONGACRE, EDWARD G., *The Man Behind the Guns* (Barnes, Cranbury, NJ, 1977). A biography of Union artillery maestro, Henry J. Hunt.

LORD, FRANCIS A., *They Fought For The Union* (Bonanza, New York 1960). A well assembled and excellently illustrated collection of facts and figures for the Union army, although rather patchy in coverage.

LORD, FRANCIS A., *Civil War Collector's Encyclopaedia* (Stackpole, Harrisburg, Pa, 1963). Somewhat less depth but greater breadth than his 1960 volume, above. This encyclopaedia covers Confederate as well as Union equipment.

LORD, WALTER, *A Time To Stand* (Harper & Row, New York, 1961). The Alamo without the mythology.

LUBBERS, HENRY G., 'Civil War Artillery', in *Wargamer's Digest*, December 1983, pp. 30–3. A useful short summary.

LUVAAS, JAY, *The Military Legacy of the Civil War* (Chicago University Press, Chicago, 1959). Exemplary scholarly analysis of the American impact on Europe during and after the war, but not of the pre-war balance of influences.

McDONOUGH, JAMES L., *Stones River – Bloody Winter in Tennessee* (University of Tennessee Press, 1980). The Battle of Murfreesboro, 1862.

McDONOUGH, JAMES L., and CONNELLY, THOMAS L., *Five Tragic Hours* (University of Tennessee Press, 1983). The battle of Franklin, 1864.

McINTYRE, BENJAMIN F., *Federals on the Frontier*, ed. Nannie M. Tilley (University of Texas Press, Austin, 1963). Memoirs from 19th Iowa Infantry.

McKIM, RANDOLPH H., *Soldier's Recollections* (Longman Green, New York, 1910; reprinted Zenger, Washington DC, 1983). Memoirs from the Confederate First Maryland Infantry.

McMURRY, RICHARD M., *John Bell Hood and the War for Southern Independence* (University of Kentucky Press, 1982). A balanced modern biography of Hood.

McWHINEY, GRADY, '"Who Whipped Whom?" – Confederate Defeat Re-examined', in *Battles Lost and Won*, ed. John T. Hubbell (Greenwood Press, Westport, Conn., 1975), pp. 261–82. An early sketch of some of the arguments developed in McWhiney and Jamieson. Well presented, stimulating but ultimately misleading.

McWHINEY, GRADY, and JAMIESON, PERRY D., *Attack and Die* (University of Alabama Press, 1982). Controversial amalgam of the findings of two highly informed students of the period.

MAHON, JOHN K., 'Civil War Infantry Assault Tactics', in *Military Analysis of the Civil War* (KTO Press, Millwood, NY, 1977), p. 253 ff; first published in *Military Affairs*, vol. 25, summer 1961, p. 57–68. This hard-hitting little article represents the 'state of the art' in Civil War tactical studies, as it is generally perceived. I hope that the present volume will change that perception.

MANNING, JAMES H., *Albany Zouave Cadets* (Privately published, New York, 1910). Regimental history of a most unusual 'regiment'.

MALONE, BARTLETT Y., *Whipt 'Em Everytime* (Chapel Hill, NC, 1919); new edn edited by W. W. Pierson jr (McCowat-Mercer, Jackson, Tenn., 1960). Memoirs from 6th North Carolina; interesting for phonetic spelling, weather bulletins and Yankee atrocities.

MILLER, REX, *Croxton's Raid* (The Old Army Press, Fort Collins, Colo, 1979). Local military history at its best, describing part of a massive Union *chevauchée* into Alabama in 1865.

MITCHELL, JOSEPH B., *Decisive Battles of the Civil War* (first published 1955; Premier Civil War Classics edn, Fawcett, New York, 1962). Altogether the finest short outline of the war that one could hope to find.

MONROE, JAMES, *The Company Drill of the Infantry of the Line* (Van Nostrand, New York, 1862). An abridgement of Hardee for 22nd New York Militia.

MOORE, J. G., 'Mobility and Strategy in the Civil War', in *Military Analysis of the Civil War* (KTO Press, Millwood, NY, 1977), pp. 106–14. The 'vital statistics' on logistics, by a meteorologist.

MORRISON, JAMES L., 'The Struggle Between Sectionalism and Nationalism at Ante-Bellum West Point', in *Battles Lost and Won*, ed. John T. Hubbell (Greenwood Press, Westport, Conn., 1975), pp. 19–30. A fine analysis of an important subject.

MOSELEY, THOMAS V., 'The Evolution of American Civil War Infantry Tactics' (unpublished doctoral thesis for University of North Carolina, Chapel Hill, 1967). Despite a very slow start, this work is probably still the best academic treatment of the subject.

NAISAWALD, L. VAN LOAN, *Grape and Canister – The Story of the Field Artillery of the Army of the Potomac* (Oxford University Press, 1960). An altogether admirable study, slightly spoiled by an unduly doctrinaire belief in the power of the rifle musket.

NISBET, JAMES C., *Four Years on the Firing Line*, ed. Bell I. Wiley (McCowat-Mercer, Jackson, Tenn. 1963). Invaluable memoirs by the commander of 66th Georgia Infantry, including his earlier experiences in 21st Georgia.

OR: *The War of the Rebellion: A Compilation of the Official Records of the Union and Confederate Armies* (US War Department, Washington DC, 1880–1901). Many commentators have pointed out that this collection contains innumerable vital references to tactics, but few have stressed that the 128 heavy volumes amount to a research project of many years' duration. I have not used them because I could not have done so systematically in the time at my disposal. It is an open question whether other serious students of Civil War tactics have done so as systematically as the case demands, but if they have not it is high time someone did.

PARISH, PETER J., *The American Civil War* (Holmes & Meier, New York, 1975). Very full modern account of the political background, pausing but rarely at the military developments.

PATRICK, ROBERT, *Reluctant Rebel*, ed. F. Jay Taylor (Louisiana State University Press, Baton Rouge, 1959). Memoirs from a member of 4th Louisiana Infantry who somehow never could avoid those clerking desk jobs, no matter how hard he tried.

POE, C., and SEYMOUR, B. (eds), *True Tales of the South at War* (University of North Carolina Press, Chapel Hill, 1961). A collection of snippets from the fighting line and the home front.

POLLARD, E. A., *Southern History of the Civil War* (first published Richmond, Va, 1862-5; reissued Blue and Gray Press, New York, n.d. (about 1961?)). The true voice of the South at War, but including many understandable distortions.

PORTER, H., *Campaigning With Grant* (Century, New York, 1897). Another solid staff history from the Wilderness campaign.

PRESSLY, THOMAS J., *Americans Interpret Their Civil War* (Princeton University Press, 1954). Good analysis of the historiographical trends.

PULLEN, JOHN J., *The Twentieth Maine* (Lippincott, Philadelphia, Pa, 1957). To my mind, one of the very best military books ever written.

REILLY, R., *The British at the Gates* (Cassell, London, 1974). Good general history of the New Orleans campaign, 1815, including healthy revisionism of the 'long rifle' myth.

REILLY, R. M., *United States Military Small Arms 1816-65* (Eagle Press, New York, 1970). An extensive listing with useful bibliography.

ROGERS, H. C. B., *The Confederates and Federals at War* (Ian Allan, London, 1973). A very good compact analysis of the various branches of the two armies.

ROSS, STEVEN, *From Flintlock to Rifle* (Associated University Presses, London, 1979). A summary of tactics, 1740-1866, somewhat wedded to the 'formation' and 'rifle' explanations.

SCOTT, WINFIELD, *Infantry Tactics, or Rules for the Exercise and Manoeuvres of the US Infantry* (Dearborn, for War Department, 3 vols, 1835). The 'baseline' for all Civil War infantry drill. A further edition was published in 1847.

SEARS, S. W., *Landscape Turned Red* (Ticknor & Fields, New York, 1983). The Battle of Antietam, 1862.

SHANNON, F. A., *The Organization and Administration of the Union Army* (2 vols, Clark, Kansas, 1928; reprinted Peter Smith, Gloucester, Mass., 1965). A refreshingly iconoclastic analysis, now rather dated - see, e.g. Carl Davis, *Arming the Union*.

SHERMAN, WILLIAM T., *Memoirs of General William T. Sherman By Himself* (new edn, 2 vols, King, London 1875). Vol. 2 discusses the operations 1864-5. The conclusion is required reading for organisation and tactics.

SHY, JOHN, 'The American Military Experience: History and Learning', in *Journal of Interdisciplinary History*, vol. 1, no. 2, 1971, pp. 205-28. A brilliant summary, laying much-needed emphasis upon the effects of past lessons for future doctrines.

SMITH, DONALD L., *The Twenty Fourth Michigan of the Iron Brigade* (Stackpole, Harrisburg, Pa, 1962). An élite regiment from Detroit.

SMITH, GUSTAVUS W., *The Battle of Seven Pines* (Crawford, New York, 1891). Exemplary tactical history at the lower levels, but unhelpful mud-slinging when it comes to the higher command. Smith was not the best friend of James Longstreet, it seems.

SMITH, MERRITT R., *Harpers Ferry Armory and the New Technology - The Challenge of*

Change (Cornell University Press, Ithica, NY, 1977). Analysis of institutional resistance to modernisation by an artisanal community.

SORREL, G. MOXLEY, *Recollections of a Confederate Staff Officer* (Neale, New York, 1905). Useful memoirs of a young man who starts as a socialite but ends as a tactician.

SPALDING, BRANCH, 'Jackson's Fredericksburg Tactics', in *Military Analysis of the Civil War* (KTO Press, Millwood, NY, 1977), pp. 67-9. An intriguing suggestion that Jackson was using very modern 'depth defence' tactics.

STARR, STEPHEN Z., 'Cold Steel: The Saber and the Union Cavalry', in *Battles Lost and Won*, ed. John T. Hubbell (Greenwood Press, Westport, Conn., 1975), pp. 107-22. Refreshing objections to the generalised contempt for the *arme blanche*.

STARR, STEPHEN Z., *The Union Cavalry in the Civil War* (Louisiana State University Press, 3 vols, 1979-85). An inspiringly monumental and well written analysis of a neglected subject. Unfortunately I failed to gain sight of Vol. 3 by the time of going to press.

STONEBRAKER, J. R., *A Rebel of '61* (Wynkoop, Hallenbeck & Crawford, New York, 1899). Memoirs from Confederate 1st Maryland Cavalry, including 'the last man killed in the Army of Northern Virginia'.

STRACHAN, HEW., *From Waterloo to Balaclava* (Cambridge University Press, 1985). Tactical thought in the British Army, 1815-54.

TANCIG, W. J., *Confederate Military Land Units* (Yoseloff, South Brunswick, NY, 1967). Just a list, but it goes down as far as company names and is therefore unique.

Thirty-fifth Massachusetts: History of the 35th Massachusetts Volunteers, 1862-5, by a committee of the regimental association (Mills Knight, Boston, 1884). Very detailed, as those from Boston tend to be.

THOMAS, D. S., *Ready - Aim - Fire!*, ed. S. V. Ash (Osborn Printing Co., Biglerville, Pa, 1981). A magnificent piece of military archaeology, analysing the bullets discovered on the Gettysburg battlefield.

THOMPSON, G. R., 'Civil War Signals', in *Military Analysis of the Civil War* (KTO Press, Millwood, NY, 1977), pp. 225-38. Excellent account of the trials and tribulations of the field telegraph.

TOCQUEVILLE, ALEXIS DE, *Democracy in America*, translated by George Lawrence (Fontana edn, London, 2 vols. 1968). Penetrating French view of America which appeared 1835-40. Part 3 of Vol. 2 gives hints on how the Americans might adapt the French approach to war for their own use.

TODD, F. P., *American Military Equipage 1851-72* (2 vols, The Company of Military Historians, Providence, RI, 1977). Uniforms and equipment, with order of battle data on the regular army.

TURBIANA, E., 'La Nostalgie dans les Armées de la Révolution' (unpublished thesis for doctorate in Medicine, *Diplôme d'État*, Paris, 1958). Very revealing study of the epidemics of homesickness which can sweep through improvised mass armies.

WAGNER, ARTHUR L., *Organisation and Tactics* (Westermann, New York, 1895). A regular officer draws lessons for the future from the wars of the nineteenth century. His framework and examples from the Civil War have been excessively borrowed by modern students in a hurry.

WARD, JOSEPH R. jr, *39th Illinois Volunteers; An Enlisted Soldier's View of the Civil War*, ed. D. D. Cummings and D. Hohweiler (Belle Publications, West Lafayette, Ind., 1981). A good account of an uneventful war spent in the siege of Charleston.

WATKINS, SAM R., *Co Aytch*, new edn, introduced by R. P. Basler (Collier, New York, 1962). Pungent memoirs of a Tennessee Confederate, written in 1881.

WEINERT, R. P., 'The Confederate Regular Army', in *Military Analysis of the Civil War* (KTO Press, Millwood, NY, 1977), pp. 16-26. Details of just how few regulars the South bothered to muster.

WELLER, JAC, 'Shooting Confederate Infantry Arms', in *The American Rifleman*, April, May and June 1954. Well researched detail of ballistics and accuracy in 'average' hands for all the major Civil War long arms.

WEST, DEAN, 'In Defense of "Johnny Reb"', in *The Courier*, vol. 6, no. 1, January-February 1985, pp.35–9. Information on the size of infantry regiments during the war, refining those in Mahon, p. 60.

WHEELER, RICHARD, *Voices of the Civil War* (Crowell, New York, 1976). Useful and well chosen 'tactical snippets'.

WILEY, BELL I., *The Life of Billy Yank* (Louisiana State University Press, Baton Rouge, 1952; reissued 1971, 1978). Marvellous distillation of 'tactical snippets' – and social snippets, even more – based on no less than 20,000 manuscript letters. Surely the biggest data base for a work of this kind.

WILEY, BELL I., *The Life of Johnny Reb* (Bobbs-Merrill, Indianapolis 1943). Classic 'snippeting' from some 10,000 manuscript letters.

WILSON, EDMUND, *Patriotic Gore* (Deutsch, London, 1962). Literary criticism of a wide range of Civil War writings, with pen portraits of some leading soldiers and musings on 'the Imperial Republic'.

WISE, GEORGE, *History of the 17th Virginia Infantry* (Kelly Piet, Baltimore, Md, 1870). A somewhat journalistic treatment of a hard-fighting unit.

WISE, JENNINGS C., *The Long Arm of Lee* (Lynchburg, 1915; reissued with introduction by L. Van Loan Naisawald, Oxford University Press, New York, 1959). Classic gunner history, written from a Confederate viewpoint with an energy and understanding that we seem to have lost.

WORSHAM, J.H., *One of Jackson's Foot Cavalry*, ed. J.I. Robertson jr (McCowat-Mercer, Jackson, Tenn., 1964). Memoirs from 21st Virginia.

ZIMMERMANN, RICHARD J., *Unit Organizations of the American Civil War* (RAFM, Cambridge, Ontario, 1982). A short and far from exhaustive treatment.

INDEX

Abattis, 63, 122, 127–8, 132
Abercrombie, 158
Adairsville, 162
Adams, Michael, 87
Air power, 53, 68, 183
Airborne infantry, 186
Alabama, CSS, 9
Alamo, 124–5
Albany Zouaves, 102
Alexander, 71, 77, 168, 177
Algeria, 24, 102, 114, 145
Amateurs, 195
Ambush, 69, 113, 142, 185
American War of Independence, 24, 27, 117
Amiens, 199
Ammunition, expenditure, 84–5, 112, 139, 155, 159, 187–89; pouches, 81–2; supply, 62, 75, 80–3, 87, 90, 105, 149, 150, 163; types, 74–6, 78, 82–6, 188
Anaconda Plan, 32
Andersonville, 192
Animal Farm, 125
Antietam, 19, 36, 48, 57, 65, 71, 78, 92, 106, 108, 110–11, 113, 118, 121, 127, 138, 141, 149, 161, 167–8, 173, 177, 179
Apaches, 9
Appomattox, 9, 23, 49, 51, 57, 65, 115, 186–88, 190
Ardant du Picq, 162, 196
Arkansas, 156
Arlington Drill School, 103, 161
Arme blanche, see Cold steel
Arnold, Jim, 196
Articulation, 106–8, 114, 123, 190
Artillery, 20, 26, 54, 58, 65–6, 84–5, 103, 118–9, 121, 123, 128–9, 157, 165–79 *passim*, 182, 185, 189; charge, 123, 172, 175–78
Atlanta, 19, 41, 47–51, 56, 59, 68, 77, 122, 127, 130, 138, 157, 171
Atrocities, 9, 18–19, 192
Attaque à outrance, 189

Attrition, 20, 23, 39, 138, 140–1
Austerlitz, 41, 46, 199
Austria, 52, 199

Bakers, 92
Balaclava, 180
Balloons, 25, 53, 68
Baltimore, 160
Bands, 91, 102
Barron, 185
Barton, Michael, 195
Battle inoculation, 87, 91, 114
Battlegroups, 65
Battle-handling, 60–7, 71, 191
Bayonet, 27, 48, 83, 100, 109, 126, 140–42, 155, 161, 165, 171, 179, 184, 195
Bazaine, 21
Beardslee, 70
Beauregard, 124–5
Beaver Dam Creek, 33, 38, 109, 156
Belgian rifle, 77, 80
Belgians, 94
Belmont, 156
Bentonville, 49, 50
Bermuda Hundred, 25
Bertrand, 99
Birney, 113
Bismarck, 192
Black Prince, 183
Black troops, 66, 101
Blankets, 82, 92, 160
Blenheim, 46
Blockade, *see* Naval aspects
Bloody angle, Murfreesboro, 139
Bloody angle, Spotsylvania, 131
Bloody Lane, 19, 149
Blue Ridge Mountains, 87, 119
Bluff, 43, 150
Bonapartism, 99
Borodino, 19, 187
Boston, 41, 78, 194
Bottles, 59, 92
Boulogne, 41
Bounty-jumpers, 94, 131–2
Bowie knife, 27, 92, 127

Bragg, 42, 44–7, 50, 65, 110, 122, 165, 198
Brandy Station, 182, 185
Breakfast, 110
Breechloading rifles, 26–7, 75–90 *passim*, 103
Brenier, 99
Brigadiers, 55, 63, 100, 103, 108, 190
Britain, 20, 27, 31, 94, 105, 125, 141, 146, 159, 180; *see also* Celts
Brooklyn, 14th Regiment, 113
Brown Bess musket, 146–7, 150
Bruises, 84
Buck 'n' ball, 74
Buckshot, 74, 86
Buddy groups, *see* Comrades in battle
Bugeaud, 141
Bugles, 58
Bull, 107
Bull Run, *see* Manassas
Bullock, 114
'Bummers', 161
Bunker Hill, 124–5
Burnside, 36–7, 57, 61, 65–6, 93, 153, 165, 191
Burnside carbine, 75
Butler, 40

Cadres, 30, 96, 98, 114, 125; *see also* Officer ratio
California, 23
Camouflage, 16, 132
Campfires, 69
Camps, 25, 69, 88, 98
Canister, 168–72, 175
Caribbean, 77
Carolinas, 41, 49, 93
Carruth, 107
Cartridges, 86, 89–90, 113, 120
Casey, 63, 102–4, 114
Casualties, 10, 16–20, 27, 32–3, 35–7, 40, 44–5, 49, 50, 84–5, 88, 91–2, 94–5, 107, 125, 127, 132, 134, 139–43, 150, 170–75, 179–80, 187, 190, 197
Cavalry, 24, 27, 38, 44–5, 49, 58, 65–7, 69–70, 80–1, 91, 99, 101, 104, 110, 121, 123–4, 176, 179–89 *passim*, 192
Cedar Creek, 175
Cedar Mountain, 179–81
Celts, 18, 35, 163
Centerville, 38
Chamberlain, 57, 120, 143

Champion's Hill, 139
Chanal, de, 98
Chancellorsville, 9, 19, 29, 37, 39, 45, 57–8, 69–70, 107, 119, 166–7, 173, 180–1, 183, 198
Chassepot rifle, 148
Chasseurs à pied, 100–2, 110, 115, 154, 161, 163, 184, 190
Cattahoochie River, 48
Chattanooga, 45–7, 50, 68, 77, 122, 129–30, 158, 178
Cheese, 161
Chew, 176
Chicago Zouave cadets, 101–2
Chickahominy River, 33, 37, 118
Chickamauga, 46, 48, 56, 65, 122, 161
Chinn house, 144
Cholera belt, 24
Ciphers, 25, 71
Circus, 102
Clausewitz, 29, 63, 198
Coe, 162
Coffee, 68, 82
Cold Harbor, 19, 39, 50–1, 132, 149, 152–3, 170, 172
Cold steel, 27, 83, 137, 143, 184
Cologne, 183
Colt revolvers, 26, 157, 181, 185
Columbia, 19
Columns, 63, 99–100, 103, 112, 151–55
Command and control, 53, 60, 62, 64, 67, 71–3, 93, 113, 119, 143, 191
Commercialism, 26
Communards, 192
Comrades in battle, 93, 101, 104, 109–10
Concentration, 62, 66, 166–68, 171, 176, 182–3
Conscription, *see* Draft
Corduroyed roads, 79, 120, 123
Corinth, 43–4, 52, 55, 58, 63, 109, 110, 113
Corn Exchange Regiment, 85
Corps, I, 5; V, 57; IX, 66, 93; XI, 69
Corps de chasse, 46, 65, 191
Corsica, 191
Cotton, 9, 32, 159
Couch, 63
Couriers, 58–9, 70–1
Court martial, 92
Cousins, 183
Cox, 130

Crane, Stephen, 10, 16-17, 20
Crater, Battle of the, 40, 66
Crimean War, 102, 125, 141
Cross-fire, 60-1, 166
Crows, 19
Cuba, 123
Cuirassiers, 67, 187
Culloden, 18
Culpeper, 38
Culp's Hill, 159; Plantation, 82
Cumberland River, 43
Curial, 99
Custer, 187

Davis, 31, 54, 191
Defiles, 100
Denmark, 52
Density of troops, 40, 55, 89, 99–105, 110, 129
Desertion, 39, 94, 110, 176
Devil's Den, 19
Dinwiddie Court House, 186, 188
Disease, 92, 94
Ditches, 127–8
Division, 4th, of IX Corps, 66
Doctrine, 9, 23, 57, 62, 67, 97, 100, 115, 126–7, 133, 145, 151, 163, 167–8, 180–82, 189–90, 192, 194
Dog tags, 50
Doubleday, 108
Double quick time, *see* Gymnastic pace
Draft, 39, 93–4, 96, 98
Drill, 22, 43, 62, 86-115 *passim*, 143-45, 150, 162, 194
Drill books, 21, 85-6, 98-108, 111-2, 114–5, 126, 141, 162-3, 189, 193, 195
Drums, 99
Duelling, 69, 87
Duels between artillery, 167–8, 174
Dumping equipment, 82–3, 86, 92, 160–1
Dust, 120

Early, 165
Earplugs, 159
Eastern Theatre and Easterners, 24-5, 33-41, 46-7, 50-1, 57, 70, 76, 87, 98, 118-21, 129, 150, 177, 191
Edgerton, 168
Egypt, 31
Elbows, 89, 155
Elections, 96

Elephant, seeing the, 10, 50
Elite units, 65, 102, 107
Ellis, 86
Ellsworth, 101–2
Enfield rifle, 26, 73, 76–9, 81, 114, 147–8, 194
Engineers, 124–27, 133, 140–1, 144, 151, 181, 192
England, *see* Britain
Erlon, d', 53
Esprit de corps, 106, 108–10
Estimating range, 88
Europe, 10, 19-22, 24, 32, 42, 52, 62, 75, 77, 79–80, 87, 96–7, 103, 111, 114-5, 126, 130, 152, 154, 162, 167, 180–82, 191, 193
Executions, 105, 110
Explosive bullets, 25
Eylau, 124

Ferrero, 66
Field glasses, 69
Fire control, 71, 89, 112–3, 141
Firefights, 16–17, 64, 83, 87, 108, 119, 131, 137–65 *passim*
First World War, 15, 20–1, 25, 40, 52, 64, 71–2, 74, 154, 194, 198–200
Fitness, 92, 101–2
Five Forks, 57, 186, 188
Flags, 25, 58–9, 71, 99, 104, 109
Flanks, 31, 37, 47–8, 60–62, 68, 104, 106, 114, 130, 186–7
Flintlock weapons, 73, 75, 78, 90, 120
Flood, 160
Foch, 198
Food, 19, 82, 97, 110, 159, 160
Footwear, 20, 79, 92
Forest, De, 16, 98, 109, 114, 143, 155, 162
Formations, 151–56, 190, 193
Forrest, 23, 44, 62
Fort Donelson, 41, 43–5, 122
Fortification, 16, 31, 33, 36, 38–40, 44-5, 47, 61, 63, 123–40 *passim*, 145–6, 149, 159, 177, 187, 189, 191
Foy, 176
France, 10-11, 21-2, 40, 52, 61, 72, 87, 93, 98–102, 105, 110, 114-5, 124–26, 141, 145, 153, 162–3, 166, 171, 176, 190–1, 193
Franco-Prussian War, 148, 191–2
Franklin, 48

Fratt, Steve, 196
Frayser's Farm, 162
Frederick II, 31, 61, 99, 187
Fredericksburg, 19, 29, 36–7, 40, 47,
    57, 65, 70–1, 77, 107–8, 118, 138,
    153, 155, 157–8, 160, 167–70,
    173–4, 177, 179
Friedland, 176
Front Royal, 38
Frontal assault, 16, 21, 33, 35–6, 40,
    46–48, 61, 131–2, 140–45, 158–63
Frying pans, 160

Gaines's Mill, 33, 48, 65, 79, 142, 155,
    179, 181
Galloway, 84
Garibaldini, 154
Gas, 25
Gatling, 192
Geary's brigade, 84
Geometry, 99, 151, 193
Georgia, 19, 49, 61, 123, 178;
    21st Regiment, 110; 66th Regiment,
    91, 129
Germany, 21, 24, 56–8, 93–4, 97,
    99, 104, 115, 124, 134, 148, 154, 157,
    183, 193, 199
Gettysburg, 19, 29, 38–9, 50, 52,
    55–6, 61, 66–7, 71, 76–7, 83–6,
    88, 94, 110, 118, 121, 153, 158–9,
    162, 166, 168–9, 174, 177, 180–1, 183
Gibbon, 113
Goldsboro, 122
Gordon, 141
Grant, 19–21, 23, 39, 43–9, 51, 54,
    57–61, 70–1, 77, 110, 121, 127, 153,
    156, 169, 186, 198
Gravelotte, 148
Green, 79, 129, 157
Green Brier, 83
Greenhorns, 32, 50–1, 87, 93, 95,
    133, 143–4, 178
Greensboro, 49
Grenades, 25
Griffin, 149, 157
Ground scouts, 182
Groundsheet, 24
Groveton, 175
Guesswork, 68, 95
'Gumbo', 120
Gustavus Adolphus, 124
Gymnastic pace, 100–1, 104, 162–3,
    184

Hagerman, Edward, 127
Haig, 198
Hair singed, 89
Hall, 71
Hall rifle, 75, 77
Halleck, 44, 52, 59, 82, 97, 124–5, 127,
    130, 152, 154, 162–3
Hamilton, 56, 58, 63
Hancock, 113, 157, 172
Hardee, 100–1, 103–4, 108, 110, 114,
    115, 127, 154, 162–3, 184, 189, 190,
    194
Harper's Ferry, 78
Haskell, 123, 176
Hats, 43, 59, 92
Haversacks, *see* Packs
Hazel Grove, 167
Headquarters, 55, 58, 67, 71, 91
Henry rifle, 75, 79, 85
Hill, A. P., 142
Hill, D.H., 63, 111, 161
Hindman, 88–9
Hiroshima, 183
Holland, 158
Hood, 19, 48, 50–1, 59, 78, 134,
    142–3, 145, 155, 162, 178, 191
Hooker, 37–9, 45, 57–8, 70, 139, 198
Hoosiers, 160
Horror, 16, 18, 20–1, 32
Horses, 175, 181–2, 184–5, 191
Horseshoes, 20
Hospitals, 92
Hoteliers, 101
Hunt, 66, 165–69
Hunting, 87, 155

Identification, 58–9, 70, 158
Illinois, 36th Regiment, 91; 55th
    Regiment, 112
Imperial Guard, 65–6
Impersonality, 16–18, 20
India, 114
Indian Rush, 154–56, 163, 190
Indiana, 14th Regiment, 78, 83, 91
Indians and Indian frontier, 96, 114,
    154–56, 160, 163
Industrial aspects, 20, 23–4, 32, 79, 81,
    90
Inherent military probability, *see*
    Guesswork
Institutional continuity, 93–4
Intelligence, 67–72
Iowa, 19th Regiment, 105, 121

Ireland, 94, 161; *see also* Celts
Iron Brigade, 102, 138, 173
Irregular warfare, 19, 95; *see also* Raids
Italy, 31, 52, 102, 141, 154

Jackson, O., 125
Jackson, T.J., 35, 37, 120, 155, 158, 175–77, 180, 186
Jamieson, Perry D., 18, 132
Japan, 31
Jena, 46
Jetersville Station, 187–8
Jogging, *see* Gymnastic pace
Johnson, 106
Johnston, A.S., 44
Johnston, J.E., 33, 47–8, 51–2, 59, 68, 127, 130, 140
Jomini, 22, 62, 99, 124–5, 151, 153
Jones, 31
Jonesboro, 129, 132
Jug handles, 61

Kearney, 63
Kenesaw, 47, 50, 84, 122, 131, 138, 149, 153, 157, 169
Kentucky, 44, 48, 93, 118, 122; rifle, 27, 75
Kernstown, 133
Kerr rifle, 74, 175
Kilpatrick, 180
Knapsacks, *see* Packs
Knoxville, 25, 82, 122
Koontz, John, 196
Korean War, 148
Ku Klux Klan, 62

Labourers, 94–5
Lancashire, 9
Language, 68, 115, 133, 194–5
Law, 142
Leadership, *see* Officer quality
Learning process, 10, 29–52 *passim*, 56–7, 66, 72, 86–7, 96, 98, 105, 124–27, 145
Lee, 9, 21–3, 29, 33, 35–9, 42, 47–9, 51, 56–8, 61, 67–8, 71, 77, 85, 98, 110, 127, 130–1, 158, 165, 171, 177–8, 186–7, 191, 198
Leipzig, 19
Liddell Hart, 183
Ligny, 134
Lincoln, 26, 32, 54, 191

Line of battle, 60, 99–101, 106, 109, 111–3, 143, 152, 185
Little Round Top, 71, 118
Livermore, 95, 197–200
Liverpool, 9
Loading sequence, 85–6, 89, 99
Logistics, 18, 32, 41, 45, 48, 54, 67, 123, 190
London, 183
Longstreet, 48, 61, 65, 83, 120, 173
Lookout Mountain, 122, 158; Valley, 158
Looting, 41, 69, 161, 179, 183
Lorenz rifle, 77
Louisiana Volunteers, 126

McClellan, 22, 32–3, 36, 38–9, 45, 52, 57, 65, 68, 70, 97, 100, 127, 169, 177, 182
McDowell, 31, 33
Machine guns, 20–1, 26, 74, 81, 192
Machinists, 94
McIntyre, 121
McKean, 113
McWhiney, Grady, 18
'Mad minute', 74
Madison Avenue, 115
Magazine rifle, 74
Mahan, 22, 124, 125, 135, 181
Maine, 1st Heavy Artillery Regiment, 174; 20th Regiment, 83, 92, 94
Malone, 158
Malvern Hill, 33, 38, 66, 118, 121, 160, 167, 170, 177
Manassas, First, 27, 30–1, 33, 36, 61, 71, 89, 97, 102, 127, 173; Second, 19, 35, 37, 48, 56, 61, 68, 83, 118, 121, 138, 144, 173–4
Maps, 58–9
Marching speeds, 99–101
Marengo, 199
Marmont, 153
Mars la Tour, 148
Marye's Heights, 65, 131, 138, 149, 160
Maryland, 36, 41, 120
Masked batteries, 31, 127
Mason-Dixon Line, 111
Massachusets, 87; 2nd Regiment, 107; 13th Regiment, 78, 86, 88, 92–3, 108, 155; 22nd Regiment, 155, 160; 35th Regiment, 78, 82, 88, 106–7, 131–2, 148, 160–1
Massaponax Creek, 108

Meade, 29, 38–9, 52, 55, 57, 59, 61, 84, 110, 133, 165
Mechanicsville, 174
Media perceptions, 16, 20, 25, 27, 68, 134
Mercenaries, 22, 99
Mexico, 21, 23, 30, 96, 100, 114, 125, 141, 145, 162
Michigan, 4th Regiment, 155; 24th Regiment, 87–8, 108
Military history, 193–96
Milledge, 84
Mine Run, 38, 50, 61, 106, 131
Mines, 25, 66
Minié, French officer and bullets, 74, 82, 114, 147, 149
Minnesota, 1st Regiment, 75, 174
Misfires, 85–6, 99
Missionary Ridge, 9, 46–7, 50, 122, 157
Mississippi, 160; Army of, 55; rifle, 77–9; River, 32, 43–4, 47, 118, 122
Missouri, 18th Regiment, 94, 113
Mist, 59, 70–1, 157–8, 185
Mixed order, 152
Mobilisation, 30, 32, 95, 191
Moltke, von, 21, 42, 97
Monroe Doctrine, 21
Mons, 74
Morale, 30–52, 93, 95, 97, 104, 143–45, 148, 162, 178
Morand, 153
Morgan, 44
Mortars, 25, 169
Mountains, 92, 118, 122
Mud, 16, 37, 61, 120, 123, 186
Mules, 158
Murat, 57, 121, 180
Murfreesboro, 45–6, 107, 122, 139
Myer, 70–1

Nansouty, 121, 180
Napoleon III, 32–3, 170
Napoleon field gun, 170–72
Nashville, 45, 48, 51, 122, 178
Naval aspects, 15, 23, 25, 43–4, 49, 77, 94, 169, 191
New Brunswick, 94
New Englanders, 60, 160
New Hope Church, 128–9
New Orleans, 41, 44, 122, 124–5, 159
New York, 70, 83, 87, 113; Fire Zouaves, 101–2; 5th Regiment, 109;

9th Regiment, 111; 49th Regiment, 86; 107th Regiment, 168; 123rd Regiment, 106–157
Ney, 179
Night fighting, 85, 91, 157–8
Nisbet, 110, 113, 129, 149, 157
Noises, 16, 58, 86, 88–9, 112, 135, 141, 149, 157, 159, 172
Non-battles, 38–40, 47–8, 50
North Anna River, 39
North Carolina, 41, 49, 107, 158
Northern Virginia, Army of, 33, 36, 40, 66, 77, 155, 171, 177, 182, 187
Nostalgia, 92
Nurses, 92

Oaths, 62, 89, 159
Officer quality, 22–4, 43–4, 46, 56, 59, 93–8, 102, 104, 112, 115, 130–1, 143–4, 167; ratio, 95–7, 102
Ohio, 63rd Regiment, 109; 66th Regiment, 84
Opequon, 155
Ordnance Board, 26, 81, 103; Department, 84
Organisation, 54, 66, 91–7, 100, 102, 113, 165–6, 173, 185
Orphan Brigade, 79, 107, 129, 142, 157
Orwell, George, 125
Ostriches, 59, 138
Overheated rifles, 73, 84, 86
Owen, 107

Packs, 82, 161, 163, 195
Paddle steamers, 25, 41
Palissades, 127
Parrott, 165
Passage of lines, 63–4, 67, 113, 190
Pea Ridge, 156
Peach Tree Creek, 84, 106, 120
Pegram's Battery, 174
Pelet, 153
Pelham, 176–7
Peninsula Campaign, 33, 35, 37–8, 45, 56, 58, 70–1, 142
Peninsular War, *see* Spain
Pennsylvania, 38, 41, 120; 1st Cavalry Regiment, 179–80, 188; 8th Cavalry Regiment, 180; 95th Regiment, 84
Pens, 128
Percussion caps, 73–4, 79, 120
Periscopes, 25

Perryville, 45, 122, 198
Pershing, 21
Petersburg, 20, 40, 47–50, 61, 66, 71, 120–1, 129, 130, 132, 149, 157, 169, 174, 186
Picket Lines, 68–9
Pickett, 18, 67, 153, 168, 186–88
Pikes, 27
Pinkerton, 68
Pistols, 26, 77, 80–1, 87, 92
Pittsburg Landing, 43
Plumbers, 94
Plunder, *see* Looting and Salvage
Point Lookout, 192
Police, 110
Politics, 54, 70, 96–7, 105
Polk, CS bishop and general, 65, 123, 169; CS colonel, 107
Polytechnique school, 124–5
Pope, 22, 35–6, 38, 68
Port Hudson, 45
Potomac, Army of, 19, 23, 33, 36–8, 40, 43, 57–8, 66, 68, 70, 98, 103, 131, 166–7, 169, 177, 179–82, 186; River, 36
Potsdam, 21
Potts' Battery, 174
Powder stains, 84
Prairie Grove, 88
Prisoners of war, 19, 68, 94, 157–8, 173, 187
Prussia, *see* Germany
Pseudonyms, 94
Purcell's Battery, 174
Pyrenees Mountains, 125

Quaker Guns, 69
Quaker Road, 143
Quality of fighting units, 95–8
Quartermasters, 75, 92
Quick-firing weapons, *see* Machine guns

Racoons, 87
Raids, 44–5, 47, 121, 183
Railroads, 20, 25, 40–1, 44, 46, 49,
Railroads, 20, 25–6, 40–1, 44, 46, 49, 53, 123, 179, 186, 191
Rains brothers, 25
Ramrods, 86, 89
Ranges of fire, 73–4, 88–90, 100–1, 104, 117–19, 127, 129, 138–9, 141–50, 159, 167–78, 190, 192, 194
Ranks for firing, 89, 99–104

Rapidan River, 38, 119
Rappahannock River, 36, 105; Station, 66, 152
Rates of fire, 73–4, 83–4, 87, 99, 139–40
Ratio of artillery, 169, 177–8; of attack, 130; of cavalry, 181, 184
Recruitment, 30, 39, 66, 94–7, 102
Red Badge of Courage, The, 10, 15, 17, 29, 53, 73, 91, 117, 137, 165, 179, 189, 193
Re-enlistment, 93
Regimental life, 91–7; strengths, 92–6
Religion, 105, 109
Remington Armament Firm, 26
Repeating rifles, 26, 31, 75–90 *passim*, 184–88
Replacements, 94–5
Resaca, 113, 142, 149
Reserves, 62–7, 72, 99, 104, 138, 184, 191
Results, 197
Revolutionary tactics, 17–18, 70, 75, 101–4, 154, 157, 176, 184–89
Reynolds, 165
Richardson, 139
Richmond, 19, 27, 31–3, 35, 40, 49, 54, 68, 80, 105, 115, 130, 186
Rifle musket, 26, 73–90 *passim*, 127, 189
Rifled cannon, 26, 123, 134, 167–9
Rifles, *see* Breechloading, Belgian, Chassepot, Enfield, Hall, Henry, Lorenz, Kentucky, Kerr, Magazine, Mississippi, Repeating, Sniper, Springfield, Spencer, Sharps, Whitworth, Winsor
River transportation, 41, 43
Rocks, 83
Roman legions, 124
Rommel, 57
Rosecrans, 46, 55–6, 58, 63
Rotterdam, 183
Rubber blanket, *see* Groundsheet
Russia, 24, 41, 124

Saber, 27, 181, 184–5, 189
St Cyr military school, 124
St Privat, 154
Salvage, 78–9, 85–6, 160–1
Sanders, Fort, 25
Saumur military school, 124
Savannah, 49, 122
Sayler's Creek, 187–8

Scale of attacks, 153, 180
Scots, 94; *see* also Celts
Scott, 20, 32, 97, 99–101, 112, 114, 127, 162
Seamen, 94
Sebastopol, 124
Second World War, 58, 148, 183
Selma, 183
Sénarmont, 176–7
Seringapatam, 158
Seven Days, 33, 174
Seven Pines, 33, 47, 63–4, 83, 103, 139, 146–7, 149, 162
Sharps Rifle, 75–6
Sharpsburg, 36, 107
Shelling, 26, 134, 169, 170–72, 191
Shenandoah Valley, 35, 41, 176, 182
Shepardstown, 85
Sheridan, 49, 57, 65, 134, 186–88
Sherman, 9, 20, 23, 43, 47–9, 51–2, 55, 59, 61, 68, 71, 94, 97, 110, 122, 127, 129–31, 133–4, 140, 144, 153, 156, 165, 178, 192
Shiloh, 43–6, 56–7, 59, 79, 110, 112, 122, 127
Shock, 140–45, 148, 151, 157, 182, 189 192
Shoemakers, 94
Shotgun, 77, 87, 127
Sicily, 154
Sigel, 156
Sights, 74, 88
Signals, 25, 53, 63, 67-72
Singapore, 31
Skirmishers, 63, 99–104, 106, 108, 110–12, 117, 135, 144, 152–57, 162–3, 168, 175, 189
Slaughter pens, 131, 140, 149
Slavery, 60, 87
Smith, 83, 139, 146
Smoke, 16–17, 59, 84, 89, 134, 150, 158
Smoothbore cannon, 26, 169–73
Smoothbore musket, 26, 32, 73–90 *passim*, 127, 129, 147, 149
Sniper rifles, 25, 74–5, 77, 81
Snipers, 25, 69, 74, 102, 123, 134, 147, 175
Snippets, 193–96
Snowballs, 87, 105
Solferino, 19
Somme, the, 20, 72
Songs, 9

South Carolina, 41, 49, 123; 1st Regiment, 89
Spade, 124–5, 127, 130, 134
Spain, 41, 141
Spencer rifle, 26, 31, 75, 77, 85, 187
Spies, 68
Spotsylvania, 39, 47, 51, 66, 84, 86, 103, 128, 131–2, 149, 152, 155, 157, 172, 174
Springfield rifle, 26, 73, 76, 78–9, 81, 148
Square formation, 99, 101, 104, 180
Staff, 33, 53–60, 62, 67, 69, 71–2, 97, 124, 133, 192
Stanton, 54, 70
Statistics, 197–200
Steuart, 159
Steuben, Von, 99
Stoneman, 183
Stone's River, 45
Stonewall, 102
Storm troops, 65–6, 104
Stress, 110–13
Stuart, J.E.B., 176, 183
Sugar, 68, 82
Sumter, Fort, 24
Surprise, 35, 37–8, 44, 62, 68, 106, 122, 141–2, 144
Surveillance, 68, 110
Swedes, 124

Tactical manuals, *see* Drill books
Tolavera, 147–8
Target shooting, 73, 86–91, 101, 126, 163
Tattoos, 105
Taylor, 143
Telamon, 18
Telegraph, 20, 25, 53–4, 68, 70–1, 191
Telescope, 69
Tennessee, 19, 45, 48–9, 122–3, 160; Army of, 77; 1st Regiment, 92; 1st Cavalry Regiment, 183; River, 43–5, 105, 122
Tents, 24
Terrain analysis, 68
Terror weapons, 132
Texas, 48, 78, 113, 145, 155, 160, 180, 185; 3rd Cavalry Regiment, 185
Thomas, 46–7, 165, 178
Tiger Rifles, 27
Tinkham, 89
Tobacco, 68

Tolstoy, 24
Tommy gun, 75
Tompkins's battery, 173
Torpedoes, *see* Mines and Naval aspects
Total war, 18
Trains, *see* Railroads
Trench raid, 157–59, 163, 190
Trenches, 128, 134
Trophies, 143

Uniforms, 20, 59, 102
Upton, 66, 103–4, 114–5, 152, 162
US, 3rd Artillery Regiment, 165; 5th Cavalry Regiment, 179

Valley Campaign, *see* Shenandoah Valley
Vampyres, 121
Vauban, 124, 135
Verdun, 20, 72
Vermont, 1st Cavalry Regiment, 180
Vicksburg, 19, 39–40, 44–7, 77, 85, 93 130, 139
Vietnam, 10, 148
Villantroys mortar, 167
Virginia, 18, 22, 36, 38–9, 41, 49–50, 52, 93, 119–22, 145, 150; 21st Regiment, 79, 133, 148; 17th Regiment, 92, 144, 162
Volley fire, 73–5, 85, 88, 99, 101, 112, 141, 143–4, 148, 150, 160

Wagons, 41, 83, 123, 181, 183
Wargames, 194
Warren, 38, 187
Washington Artillery of New Orleans, 107
Washington DC, 23, 31, 40, 54, 68, 70, 78, 80, 115, 127, 130
Waterloo, 18, 29, 53, 99, 119, 176, 179, 193
Watkins, 31, 84, 92, 114–5
Weapons, availability of, 76–81; costs of, 81, 181

Weather, 50, 117–22 *passim*
Webster's Dictionary, 195
Weed, 166
Wellington, 11, 29, 36, 57, 119, 125, 134, 141, 145, 158, 160
West Point, 22–3, 27, 95–7, 114, 124–26, 141, 165, 181, 191
West Virginia, 1st Cavalry Regiment, 180
Western Front, 20, 54, 154, 197–200
Western Theatre and Westerners, 23-24, 39–48, 51, 56, 61, 66, 70–1, 76–77, 82, 87, 98, 121–23, 129, 150, 178, 182–3
Wheat's Battalion, *see* Tiger Rifles
Whiskey, 59–60, 106, 112
White Oak Road, 57, 186–7
White Oak Swamp, 177
Whiting, 142
Whitworth rifle, 74, 80, 114, 175
Wig Wag, 25, 69–71
Wilderness, 19, 20, 37, 39, 47, 50–1, 57–8, 61, 82, 87, 104, 113, 119–21, 149, 160, 167, 169–71, 173, 186, 191
William II, 194
Williams bullet, 82
Wilson, 183
Winchester firearms, 26
Winchester, Virginia, 155
'Winsor' rifle, 78
Winter quarters, 37, 50, 120
Wire entanglements, 25, 160
Wire-tapping, 68
Wisconsin, 94; 6th Regiment, 113
Wise, 171, 175
Woods, 31, 37, 44–5, 58, 63, 68, 70–1, 106, 108, 111, 113, 117–27, 134, 158, 167, 175, 181
Worsham, 79

Yells, 18, 38, 89, 142–3, 159–60
Yorktown, 25, 33, 48, 52

Zouaves, 101–2, 113, 154
Zuparko, Ned, 196